Highest Duty

Highest Duty

My Search for What Really Matters

Captain Chesley "Sully" Sullenberger

with Jeffrey Zaslow

An Imprint of HarperCollinsPublishers

HarperCollins books may be purchased for educational, business, or sales promotional use. For information please write: Special Markets Department, HarperCollins Publishers, 10 East 53rd Street, New York, NY 10022.

FIRST HARPERLUXE EDITION

HarperLuxe™ is a trademark of HarperCollins Publishers

Library of Congress Cataloging-in-Publication Data is available upon request.

ISBN: 978-0-06-192758-4

09 10 11 12 13 ID/OPM 10 9 8 7 6 5 4 3 2 1

Highest Duty is dedicated to
my wife, Lorrie, and my daughters, Kate and Kelly.
You are the three most important people in my life,
and I love you more than I can express in words.

This book is also dedicated to
the passengers and crew of Flight 1549
and to their families.
We will be joined forever
because of the events of January 15, 2009,
in our hearts and in our minds.

Contents

Highest Duty

1

A FLIGHT YOU'D NEVER FORGET

The flight lasted just a few minutes, but so many of the details are rich and vivid to me.

The wind was coming from the north not the south, which was unusual for that time of year. And my wheels made a distinct rumbling sound as they rolled across the rural Texas airstrip. I remember the smell of the warm engine oil, and how it drifted into the cockpit as I prepared to take off. There was also the smell of freshly cut grass in the air.

I have a clear recollection of how my body felt—this heightened sense of alertness—as I taxied to the end of the runway, went through my checklist, and got ready to go. And I recall the moment the plane lifted into the air and, just three minutes later, how I would need to return to the runway, intensely focused on the tasks at hand.

All these memories are with me still.

A pilot can take off and land thousands of times in his life, and so much of it feels like a speeding blur. But almost always, there is a particular flight that challenges a pilot or teaches or changes him, and every sensory moment of that experience remains in his head forever.

I have had a few unforgettable flights in my life, and they continue to live in my mind, conjuring up a host of emotions and reasons for reflection. One took me to New York's Hudson River on a cold January day in 2009. But before that, perhaps the most vivid was the one I've just described: my first solo flight, late on a Saturday afternoon at a grass airstrip in Sherman, Texas. It was June 3, 1967, and I was sixteen years old.

I hold on to this one, and a handful of others, as I look back on all the forces that molded me as a boy, as a man, and as a pilot. Both in the air and on the ground, I was shaped by many powerful lessons and experiences—and many people. I am grateful for all of them. It's as if these moments from my life were deposited in a bank until I needed them. As I worked to safely land Flight 1549 in the Hudson, almost subconsciously, I drew on those experiences.

For a few months when I was four years old, I wanted to be a policeman and then a fireman. By the time I was five, however, I knew exactly what I wanted to do with my life—and that was to fly.

I never wavered once this possibility came into my head. Or more precisely, came over my head, in the form of jet fighters that crisscrossed the sky above my childhood home outside Denison, Texas.

We lived by a lake on a sparse stretch of land nine miles north of Perrin Air Force Base. Because it was such a rural area, the jets flew pretty low, at about three thousand feet, and you could always hear them coming. My dad would give me his binoculars, and I loved looking into the distance, to the horizon, wondering what was out there. It fed my wanderlust. And in the case of the jets, what was out there was even more exciting because it was coming closer and closer at a very high rate of speed.

This was the 1950s, and those machines were a lot louder than today's fighters. Still, I never came across people in my part of North Texas who minded the noise. We had won World War II not long before, and the Air Force was a source of pride. It wasn't until decades later, when residents near air bases began talking about the noise, that pilots felt the need to answer the complaints. They'd sport bumper stickers that said JET NOISE: THE SOUND OF FREEDOM.

Every aspect of airplanes was fascinating—the different sounds they made, the way they looked, the physics that allowed them to rocket through the sky, and most of all, the men who controlled them with obvious mastery.

I built my first model airplane when I was six years old. It was a replica of Charles Lindbergh's *Spirit of St. Louis.* I read a lot about "Lucky Lindy" and understood that his flight across the Atlantic wasn't really about luck. He planned. He prepared. He endured. That's what made him heroic to me.

By 1962, when I was eleven years old, I was already reading every book and magazine I could find that talked about flying. That was also the year I took my first plane ride. My mom, a first-grade teacher, invited me to accompany her to a statewide PTA convention in Austin, and it was her first plane ride, too.

The airport, Dallas Love Field, was seventy-five miles south of our house, and when we got there, it seemed like a magical place filled with larger-than-life people. Pilots. Stewardesses. Well-dressed passengers with somewhere to go.

In the terminal, I stopped at the newly installed statue of a Texas Ranger. The plaque read ONE RIOT, ONE RANGER, and told the apocryphal story of a small-town disturbance in the 1890s. A local sheriff had called for a company of rangers to stop the violence, and when only one ranger showed up, the townspeople were taken aback. They'd asked for help and now wondered if they were being denied. "How many riots do you have?" the ranger allegedly asked. "If y'all got just one, all you need is one ranger. I'll take care of it."

I also saw another hero that day at the airport. I had been enthralled by the early Project Mercury space missions, so I was excited to spot a short, thin man walking through the terminal. He was wearing a suit, a tie, a hat, and his face was completely familiar to me. I recognized him from television as Lieutenant Colonel John "Shorty" Powers, the voice of Mission Control. I couldn't bring myself to approach him, though. A guy who had all these astronauts to talk to didn't need an eleven-year-old kid tugging at his jacket.

It was a cloudy day, a little rainy, and we walked out on the tarmac to climb a staircase onto our Braniff Airways flight, a Convair 440. My mom wore white gloves and a hat. I was in a sport coat and slacks. That's how people traveled then. In their Sunday best.

Our seats were on the right side of the aircraft. My mom would have loved to look out the window, but she knew me. "You take the window seat," she said, and even before the plane had moved an inch, my face was pressed against the glass, taking everything in.

As the plane sped down the runway and began to rise, I was wide-eyed. My first thought was that everything on the ground looked like a model railroad layout. My second thought was that I wanted this life in the air.

It took a few more years for me to return to the skies. When I was sixteen, I asked my dad if I could take flying lessons. He'd been a dental surgeon in the Navy during World War II. He had great respect for aviators, and he clearly saw my passion. Through a friend, he got the name of a crop-dusting pilot named L. T. Cook Jr., who had a landing strip on his property nearby.

Before World War II, Mr. Cook had been an instructor in the federal government's Civilian Pilot Training Program. At the time, isolationists didn't want the United States getting involved in the war in Europe. But President Roosevelt knew the United States was likely to enter the conflict and would need thousands of qualified pilots. Starting in 1939, veteran fliers such as Mr. Cook were charged with training civilians so they'd be ready when and if war was declared. The program was controversial at the time, but as things turned out, all of those prepared pilots helped the Allies win the war. Mr. Cook and pilot trainers like him were the unsung stateside heroes.

When I met him, he was in his late fifties, and a no-nonsense, all-business kind of man. Most of his time was spent crop dusting, but if he saw someone who seemed to have the smarts and temperament to fly, he'd take him on as a student.

I guess he liked the look of me well enough. I was this tall, quiet, earnest kid, and I was respectful because my parents had taught me to be deferential to my elders. I was also the classic introvert, and he wasn't a guy who needed much conversation. He saw I was serious about flying and that I had an obvious enthusiasm, despite my low-key demeanor. He said he'd charge me six dollars per hour for the airplane. That was the "wet rate" because it included fuel. For his time training me, he asked for another three dollars an hour. My parents paid for the airplane, so for a thirty-minute flight I owed him just $1.50 for his instructor's fee. I paid for my share from money I earned in my job as a church janitor.

I have logbooks going back decades, covering thousands of flights. And in my first logbook, my very first entry was April 3, 1967, when Mr. Cook took me up for thirty minutes. We flew in a tandem two-seater, an Aeronca 7DC. It was a very basic propeller airplane, built in the late 1940s. It didn't even have a radio. I had the controls in my hands from pretty much the first moment.

I sat in front, Mr. Cook sat in back with his own set of controls, and he did what pilots call "following you through." That meant he'd keep his hands hovering over his stick so he could instantly take command if I went astray with my stick. He shadowed my movements, shouting directions over the noise of the engine. As so many pilots did in the early years, he used a cardboard megaphone to aim the sound of his voice right in my ear. He spoke only when he needed to, and he rarely gave a compliment. Still, in the weeks that followed, I sensed that he thought I was catching on, and had the right instincts. I studied flying at home every night, too, taking a correspondence course that prepared me for the private pilot license written exam. Mr. Cook saw I was devoted.

Sometimes I'd arrive for a lesson and he wouldn't be there. So I'd drive into town because I knew exactly where to find him: drinking coffee at the local Dairy Queen. He'd finish his coffee, toss a tip on the table, and we'd go back to his strip.

He gave me sixteen lessons over the next couple of months, each averaging thirty minutes in the air. By June 3, my total flying time added up to seven hours and twenty-five minutes. That day, he took me up for a flight, and after ten minutes of flying around, he tapped me on the shoulder.

"All right," he said. "Bring it in for a landing and taxi over to the hangar." I did as I was told, and when we got there, he hopped out of the plane. "OK," he said. "Take it up and land three times by yourself."

He didn't wish me luck. That wasn't his way. I'm not saying he was gruff or unfeeling. It's just that he was very matter-of-fact about things. He had obviously decided: The kid's ready. Let him go. He expected I wouldn't fall out of the sky. I'd be OK.

These days, a boy couldn't get into the air alone so quickly. Airplanes are more complex. There are all sorts of requirements and insurance issues that have to be taken care of before someone flies solo. The air traffic control system is more complicated. And instructors may be more protective, worried and wary, too.

But that day, in the North Texas countryside, I didn't have to deal with air traffic control or complicated regulations. It was just me and the plane, and Mr. Cook, who was watching me from the ground.

Because the wind was coming from the north, I had to go to the opposite end of the runway so I could take off in that direction. That wasn't the usual direction, but I got my bearings and prepared to go.

The strip was lower at the south end and sloped uphill toward the north. And even though Mr. Cook had just mowed his grass strip, it wasn't as smooth as a paved runway or a putting green.

Alone at the end of an airstrip for the first time in my life, I checked the ignition and the oil pressure. I made sure the engine, rudder, elevator, and ailerons were working properly. I went through everything on my checklist. And as my hand tightened on the control stick, I took a breath, released the brakes, and began my takeoff. Mr. Cook had told me that I'd be leaving the ground more quickly than I was used to. The reason? The plane was now lighter with him not in it.

When this type of airplane heads down a runway and is ready to fly, it just lifts off. But when a new pilot is ready to fly alone, someone has to say so. That someone was the laconic Mr. Cook, nodding there on the sidelines as I rose into the air while he grew smaller and smaller in the field below me. I was grateful to him.

Climbing to eight hundred feet above the ground, and then circling the field, I felt an exhilarating freedom. I also felt a certain mastery. After listening, watching, asking questions, and studying hard, I had achieved something. Here I was, alone in the air.

I don't think I was smiling about my good fortune. I was too busy concentrating to allow myself to smile. And I knew Mr. Cook was watching me from under his baseball cap, his head tilted upward. I wanted to look good for him, to do everything right. I didn't want him to have a long list of things to critique me about when I landed.

As I flew, it was as if I could hear his voice. *Use the rudder to keep the controls coordinated.* Even though he wasn't there in the airplane, his words were still with me.

I was too busy to do any sightseeing. I flew over a little pond, and the town of Sherman was off to my left. But my goal was not to enjoy the view. My goal was to do this well enough so that Mr. Cook would let me do it again.

He had instructed me to make the usual rectangular pattern around the landing strip, which took about three minutes in flight, so I could practice touching the runway, lifting back into the air, and then coming back around to do it again. I had to do this three times before coming in for a final landing.

My entire first solo experience was only nine minutes or so, but I knew it was a crucial first step. I'd done my reading: In 1903, Orville Wright's first flight had traveled a distance of forty yards, had risen twenty feet in the air, and had lasted just twelve seconds.

Mr. Cook greeted me when it was all over, and as I shut down the engine, he said I'd done what he'd asked. There was no "atta boy," but I knew I'd passed the test. He told me he'd be busy crop dusting in his other plane much of the summer, and so I might as well just keep taking his Aeronca up to practice on my own. We agreed that I could return every few days to hone my skills, alone in the sky, for six dollars per hour.

Now, at age fifty-eight, I have 19,700 hours of flying time under my belt. But I can trace my professional experience back to that afternoon. It was a turning point. Though I had less than eight hours in the air, Mr. Cook had given me confidence. He had given me permission to discover that I could get a plane safely into the air and then safely back to the ground. That first solo flight served as confirmation that this would be my livelihood, and my life.

I didn't completely focus on it at the time, but I realize now that my entrance into the world of piloting was very traditional. This is how people had learned to fly since the beginning: an older, veteran pilot teaching the basics to a youngster from a grass strip under an open sky.

I look back and appreciate very much that I was a lucky young man. It was a wonderful start.

No one else in my high school was interested in being a pilot, so I was alone in my pursuit. I had friends, but a lot of the other kids saw me as this shy, studious, serious boy always reading flight manuals and heading out to the airstrip. I was not easily outgoing. I was more comfortable in a cockpit.

In some ways, I grew up fast on that airstrip, learning things that helped me see the possibilities in life, and the great risks.

One day, when I got out to Mr. Cook's hangar, I noticed a Piper Tri-Pacer, painted white with red trim, crumpled on the field at the north end of the runway. Mr. Cook told me the story. A friend of his was bringing the Tri-Pacer in for a landing, approaching the airstrip, and he had to cross over U.S. 82. He didn't realize until it was too late that there were twenty-foot-high power lines stretched along the highway. He pulled up the nose of the plane to clear the wires, but that action caused him to slow down and lose lift. His plane slammed down nose-first into the ground, and he died instantly.

No one had come yet to collect the wrecked plane, and so there it still sat at the end of the airstrip. I walked a quarter mile up to it and looked inside at the blood-splattered cockpit. In those days, airplanes had only lap belts, not shoulder harnesses, and I figured that his head must have hit the instrument panel with great violence. I tried to visualize how it all might have happened—his effort to avoid the power lines, his loss of speed, the awful impact. I forced myself to look in the cockpit, to study it. It would have been easier to look away, but I didn't.

It was a pretty sobering moment for a sixteen-year-old, and it made quite an impression on me. I realized that flying a plane meant not making mistakes. You had to maintain control of everything. You had to look out for the wires, the birds, the trees, the fog, while monitoring everything in the cockpit. You had to be vigilant and alert. It was equally important to know what was possible and what was not. One simple mistake could mean death.

I processed all this, but that sad scene didn't give me pause. I vowed to learn all there was to know to minimize the risks.

I knew I never wanted to be a hot dog—that could get me killed—but I did make my own fun. I'd tell my parents and younger sister to step outside our home at an appointed time, and then I'd fly over and waggle the wings up and down to say hello to them. We lived in such a sparsely populated area that regulations allowed me to fly as low as five hundred feet above the house. My family couldn't exactly see my face, but they could see me waving at them.

By October 1968, after seventy hours in the air, I was ready to try for a private pilot certificate, which required a "check ride" with an FAA examiner. I passed, which allowed me to fly with a passenger.

I thought the honor of first passenger ought to go to my mother, and my logbook shows I took her for a ride on October 29, 1968, the day after I got my certificate. I put a simple little star next to the flight data in the logbook; a small acknowledgment of a special moment. It was the 1960s equivalent of an e-mailed smiley face.

My mom didn't seem nervous that day, just proud. As I helped her into the back seat and strapped her in, I described the sounds she would hear, what we'd see, how her stomach might feel. The upside of my being so serious, I guess, is that I struck people as responsible and able. I wasn't a flouter of rules. And so my mom had confidence in me. She just sat back, her life in my hands, with no urge to be a back-seat driver. She let me chauffeur her around in the sky, and when we landed, she hugged me.

The possibility of having passengers opened up a new world, and after I took my sister, my dad, and my grandparents for a ride, I found the courage to ask someone else. Her name was Carole, and she was a cute, slender girl with brown hair and glasses. We went to Denison High together, and we were also in our church choir. I had a crush on her, and I liked to think she had noticed me, too. There are girls who are good-looking and know it, and have the luxury of getting by on their beauty. Carole was attractive, yet she didn't carry herself like those girls. Even though she wasn't overtly outgoing, she had an open, friendly manner that just drew people in.

No girl had ever expressed much interest in my experiences as a pilot. This was long before the movie *Top Gun*, and in any case, I wasn't Tom Cruise. Besides, flying was an abstract thing. No one saw me doing it. It's not like I caught winning touchdown passes and had my picture in the local paper. Everything I did was out of view and high in the sky. If I mentioned flying to girls, they never seemed hugely impressed. It sometimes felt like they were bored with the conversation. Or maybe I wasn't able to find the right words to convey the majesty of it.

In any case, I decided to see if I could interest Carole. She was quiet—similar to me in that way—and so it was often difficult keeping conversations going with her. When I asked her if she'd like to go flying with me, I had no expectations. Even if she wanted to go, I figured her parents wouldn't allow it. But she asked them, and they agreed to let her go on a forty-five-minute trip across the Arkansas and Poteau rivers to Fort Smith, Arkansas.

This was my effort at a date, and I was pretty thrilled that it was going to happen. Looking back, it's remarkable that her mom and dad said yes. In essence, they were allowing a boy, not yet eighteen years old, to take their underage daughter across state lines. In a light airplane, no less.

And so we went. It was a clear, cold day with smooth air and good visibility. You could see for miles in any direction. Airplanes are noisy, so it's hard to converse. I'd yell, "That's the Red River down there," and she'd yell back, "What?" and I'd repeat myself. But I was so happy to have her aboard.

We flew in a Cessna 150 I'd rented for two hours. This was a very small airplane, with room only for two people, sitting side by side. The whole cabin was just three feet wide, and so my right leg was touching her left leg. There was no other way to do it.

Picture me, seventeen years old, with this pretty girl next to me, her leg touching mine for two hours, my arm rubbing against her arm. I could smell her perfume, or maybe it was her shampoo. Once in a while she'd lean over me to look at the sights out my window, her hair brushing against my arm. It was a new experience for me, realizing that flying could be such a sensual experience.

Was it hard for me to concentrate on the controls? No. I guess that was just another example of how a pilot has to learn to compartmentalize. I was completely aware of Carole, but I was also on task and responsible.

I wanted to woo her, but my most important job was to keep her safe.

Not much came of our relationship, but that flight—sitting so close to her, shouting out landmarks of the Texas countryside, taking her to lunch at the Fort Smith airport—well, it's just a sweet, warm memory.

A pilot can have thousands of takeoffs and landings, most of them unremarkable. Certain ones, though, he never forgets.

The last time I was out at L. T. Cook's airstrip was in the late 1970s. I had lost touch with him in the early eighties, and I later learned he had cancer, and had several tumors removed from his neck and jaw. Some people speculated that his illness was a result of all the crop-dusting chemicals he sprayed every day. He died in 2001.

After my emergency landing of US Airways Flight 1549 in the Hudson River, I got thousands of e-mails and letters from people expressing gratitude for what my crew and I did to save all 155 people on board. In one stack of mail, I was thrilled to discover a note from Mr. Cook's widow, whom I hadn't heard from in years. Her words lifted my spirits. "L.T. wouldn't be surprised," she wrote, "but he certainly would be pleased and proud."

In many ways, all my mentors, heroes, and loved ones—those who taught me and encouraged me and saw the possibilities in me—were with me in the cockpit of Flight 1549. We had lost both engines. It was a dire situation, but there were lessons people had instilled in me that served me well. Mr. Cook's lessons were a part of what guided me on that five-minute flight. He was the consummate stick-and-rudder man, and that day over New York was certainly a stick-and-rudder day.

I've done a lot of thinking since then about all the special people who mattered to me, about the hundreds of books on flying that I've studied, about the trage-dies I've witnessed again and again as a military pilot, about the adventures and setbacks in my airline career, about the romance of flying, and about the long-ago memories.

I've come to realize that my journey to the Hudson River didn't begin at LaGuardia Airport. It began de-cades before, in my childhood home, on Mr. Cook's grass airfield, in the skies over North Texas, in the California home I now share with my wife, Lorrie, and our two daughters, and on all the jets I've flown toward the horizon.

Flight 1549 wasn't just a five-minute journey. My entire life led me safely to that river.

2

A PILOT'S LIFE

was lucky enough to discover my passion for flying when I was very young, and to indulge that passion day after day. Lucky that some things went my way; my eyesight, for instance, was good enough to allow me to become a fighter pilot. And lucky that when I left the military, I found work as an airline pilot, when such jobs weren't plentiful.

I still feel fortunate, after all these years, to be able to follow my passion. The airline industry has its problems, and a lot of the issues can be troubling and wearying, but I still find purpose and satisfaction in flying.

There's a literal freedom you feel when you're at the controls, gliding above the surface of the earth, no longer bound by gravity. It's as if you're rising above the nitty-gritty details of life. Even at a few thousand feet, you get a wider perspective. Problems that loom large down below feel smaller from that height, and smaller still by the time you reach thirty-five thousand feet.

I love that flying is an intellectual challenge, and that there's mental math that needs to be done all along the way. If you change the angle of the nose versus the horizon by even one degree while traveling at a typical commercial airliner speed of seven nautical miles a minute, it's enough to increase or decrease your rate of climb or descent by seven hundred feet per minute. I enjoy keeping track of all the calculations, staying aware of the weather conditions, working with a team—flight attendants, air traffic controllers, first officers, maintenance crews—while knowing intimately what the plane can and cannot do. Even when the controls are being manipulated through automation, pilots have to back up the computer systems with their own mental math. I like the challenge of that.

I also like sharing my passion for flying. It's a disappointment to me that a lot of kids today aren't especially fascinated by flight. I've watched countless children walk past the cockpit without paying much attention; they're too focused on their video games or their iPods.

When there are children who eagerly want a look inside "my office" at the front of the plane, their enthusiasm is contagious. It's so gratifying to see their excitement about something I care deeply about. If we aren't busy during boarding, the first officer and I enjoy inviting inquisitive children to sit in our seats in the cockpit, ask questions, and let their parents take photos of them wearing a captain's hat.

Being a pilot has a tangible end result that is beneficial to society. It feels good to take a planeload of 183 people where they need or want to go. My job is to reunite people with family and friends, to send them on long-awaited vacations, to bring them to loved ones' funerals, to get them to their job interviews. By the end of a day, after piloting three or four trips, I've taken four or five hundred people safely to their destinations, and I feel as if I've accomplished something. All of them have their own stories, motivations, needs—and helping them brings a rewarding feeling.

This is what gets me ready for work, and one of the things I look forward to.

I did not kiss my wife good-bye.

It was five-thirty Monday morning, and I was leaving home for a four-day trip. My schedule had me piloting seven US Airways flights, with the last leg set for Thursday, January 15: Flight 1549 from New York to Charlotte.

I didn't kiss Lorrie because, over the years, I've come to realize that Lorrie is a light sleeper, and though I'd like to quietly kiss her before every trip and whisper "I love you," doing so at 5:30 A.M. wouldn't be fair to her. I'd leave, and she'd be left there in bed, eyes open, to contemplate everything that she and our two daughters needed to do in the days ahead—all of it without me or my help.

Despite my passion for flying, the constant departures that define a pilot's life have been very hard on us. Gone from home about eighteen days per month, I have missed well over half of my children's lives.

My leaving isn't an indication that I love flying more than I love my wife and kids. In fact, Lorrie and I have talked in recent years about my doing something besides commercial aviation, something that would keep me closer to home. Despite the limits on how a man can reinvent himself, I've been confident about finding another way of meeting my family's financial needs that would equal being an airline captain. But I've wanted it to be a good fit that would take advantage of my life experiences. In the meantime, my dedication to the profession remains strong. And Lorrie knows me.

She knows what flying means to me. We've found our ways to cope.

And so on that Monday, like so many before, I took my leave. Lorrie and our daughters, Kate, sixteen, and Kelly, fourteen, were fast asleep when I pulled the car out of our garage in Danville, California, and headed for San Francisco International Airport.

As the sun rose, I was already thirty-five miles away, crossing over San Francisco Bay on the San Mateo Bridge. I needed to be on a 7:30 A.M. flight to Charlotte—as a passenger.

Flight crews all have a base of operation, and mine is Charlotte, North Carolina. I used to be based in San Francisco, beginning in the early 1980s, when I flew for Pacific Southwest Airlines. In 1988, PSA merged with USAir, and I became a USAir pilot. In 1995, when USAir closed its San Francisco base, my base became Pittsburgh and then Charlotte. Lorrie and I wanted to remain in California, so like others based far from home, I've made a decision to commute across the country to start my work. We have chosen this life, and I'm grateful the airline allows it. Still, the logistics of it are wearying.

I don't have to pay for my flights to get to work, but I do have to go standby. If no seat is available, I can usually ride in the jumpseat in the cockpit. That's my ace in the hole. Mostly, though, I prefer to be in the back of the plane, out of the way of the pilots doing their job. In the back, I can read a book or close my eyes and try to sleep.

Because I'm in uniform, passengers will sometimes ask me a question about the flight, the turbulence, or how to best jam their overstuffed bags into the overhead compartment. Just as often, no one really notices me.

That's how it was on the flight that day to Charlotte. I sat there in my middle seat in coach, as anonymous as always, with no conception that by week's end everything would change. These were the final days of my old familiar life as a pilot.

I am a man of routine, and there's a precision to my life that leaves Lorrie rolling her eyes sometimes. She says I'm very controlled and regimented, and though she believes that is part of what makes me a good pilot, it also makes me hard to live with on occasion. Lorrie knows other pilots' spouses who describe them the same way. Like me, they'll come home after days away and try to take charge, annoying loved ones by reorganizing the dishes in the dishwasher, finding a more efficient way to stack everything. I guess the flying culture—all our training—is what makes us so organized. Or, as Lorrie suspects, maybe there's a certain type of personality attracted to the profession.

In any case, I suppose I'm guilty as charged. But my exacting approach to things may serve me well in a lot of ways.

I had packed for this four-day trip the same way I pack for every four-day trip. I never want to bring more than necessary. I wore my captain's uniform—jacket and pants—and in my pilot's "roll-aboard" carry-on, brought three clean shirts, three pairs of underwear, three pairs of socks, my shaving kit, running shoes, an umbrella, an iPod, my laptop to check e-mail, and four books to read. I also had my American Express *SkyGuide*, which lists the complete North American flight schedule for all airlines. In a shirt pocket I had a US Airways trip sheet, with a full itinerary for the four days. Since my travels would take me to Pittsburgh and New York, where the weather would be cold and possibly snowy, I also brought a long winter overcoat, gloves, and a knit cap.

I enjoy listening to music on an iPod when I am in a city for an overnight. I always try to make a point of leaving the hotel to go for a walk, with music in my ears. Lately, I've been partial to Natalie Merchant, Green Day, the Killers, and Evanescence. I also find myself listening again and again to the works of Fritz

Kreisler, the legendary Austrian violinist. He composed and recorded *Liebesleid* (Love's Sorrow) and *Liebesfreud* (Love's Joy), which is an inspiring sound track on a walk or run around a city, lost in your own thoughts.

In recent years, I've also been spending more time on the road focused on my future up the road. I am fifty-eight years old, and I face mandatory retirement from the cockpit when I turn sixty-five. What will I do then? Since September 11, 2001, the airline industry has been ailing, and as a result of cutbacks, I've lost 40 percent of my salary. Meanwhile, the US Airways pension I thought I could count on was terminated in 2004, and a government-backed replacement plan is a very weak substitute. As a result, I've lost more than two-thirds of my pension. My story is a familiar one across the airline industry.

Trying to earn money elsewhere, I've bought some real estate over the years, with mixed results. I own a property in Northern California that used to house a Jiffy Lube oil-change franchise. The operation didn't renew its lease, however, and I've been unable to find a new tenant. So as I sat on that flight to Charlotte, I went over some of those details in my head.

About a year ago, I also started my own side business, a consulting company called Safety Reliability

Methods, Inc. It seemed like the right fit for me as my flying career winds down. Long before the landing in the Hudson, I'd been passionately involved in matters of air safety, dating back to my days as an Air Force fighter pilot. And so I brought three books on this four-day trip that were related to issues I want to address as a consultant.

I've been slowly building my firm, designed to help those in other occupations benefit from the airline industry's tactical and strategic approaches to safety. Pilots have extensive checklists that we follow in the cockpit. My firm encourages initiatives, such as those now under way in medicine, that mirror pilots' checklists. For instance, the World Health Organization now suggests the use of surgical safety checklists, requiring hospital teams to make certain that a patient's known allergies are checked, and instruments, needles, and sponges are counted to make sure none are left inside a patient.

I think commercial aviation is ultrasafe. Given the number of passengers we deliver safely to their destinations each day, and the relatively low risk associated with flying, our record so far is commendable. But airline companies must remain diligent, especially in the face of all the economic cutbacks plaguing the industry, or our good record could be compromised.

One of the books I had with me on that trip was *Just Culture* by Sidney Dekker, borrowed from my local library. Dekker writes about the balancing act between accountability and learning when it comes to people reporting safety issues. I have long believed that we can make a company culture, government, or community safer by encouraging people to report their own mistakes and safety deficiencies. So this book was a confirmation of my own study of these issues and my years of experience as a pilot.

As I sat in my middle seat on the way to Charlotte, I found myself reading and taking notes for my consulting business. I don't recall trading too many words with the passengers on either side of me.

When I'm a passenger in the back of a plane, though I'm reading or trying to nap or worrying about the shuttered Jiffy Lube, I still have a general awareness of how the flight is going and what the pilots are doing. I can feel the movements of the airplane. Most of my fellow passengers are engaged with their own books or are tapping away on their laptops, and they don't realize subtle things. But even when I'm not trying, I can tell when the plane is climbing or descending, or when the pilots are changing the flap setting or the engine thrust. For pilots, that general awareness comes with the territory.

The flight I was on had left San Francisco at 7:30 A.M. Pacific time, and arrived in Charlotte at 3:15 P.M. Eastern time. I got something to eat at the airport in Charlotte and then made my way to the gate for my first piloted flight of the four-day trip. I'd be going right back to San Francisco, flying an Airbus A321, carrying about 180 passengers.

Once I got to the gate, I smiled at some of the passengers and greeted the three flight attendants—Sheila Dail, Donna Dent, and Doreen Welsh. I had flown with Sheila and Donna before. I'm guessing I had shared trips with Doreen, too, some years ago, when we were both based in Pittsburgh. Because US Airways hasn't hired new flight attendants in years, all our crews are veterans. Doreen, now fifty-eight, joined the company in 1970 when it was Allegheny Airlines. That's thirty-eight years of experience. Both Sheila, fifty-seven, and Donna, fifty-one, have more than twenty-six years with the airline.

At the gate, I also shook hands with Jeff Skiles, the first officer who'd be flying with me. He and I had never met before, so we introduced ourselves. Along with Sheila, Donna, and Doreen, we'd be a team for the next four days.

Despite all my years as a pilot, it's common for me to have a first officer or flight attendants I've never met.

Even after some serious downsizing, US Airways still has about 5,000 pilots and 6,600 flight attendants. It's impossible to know them all.

It is standard at our airline for a crew to have a brief meeting together at the start of a trip. It's vital to make individuals feel like a team quickly so that they can work almost as well together on the first flight as they naturally would after having flown several flights together. So before the passengers boarded we stood—Jeff, Sheila, Donna, Doreen, and I—in the aisle of the empty first-class cabin for a couple minutes, and I said a few words.

As the captain, it's up to me to set the tone. I want to be approachable. I asked the flight attendants to be my eyes and ears during the days ahead, to tell me about anything important that I couldn't observe from the cockpit. I asked them to let me know what they needed to do their jobs—catering, cleaning, whatever—and told them I'd try to help. I wanted them to know I was looking out for them. "I can't get you your retirement plans back, but I can do a few things that will make your quality of life better. One of them is, when we arrive at our destination on the last flight of a day, I'll call the hotel and make sure that they've sent the van so we're not waiting for twenty minutes."

Jeff, forty-nine years old, was very friendly from the moment we said hello, and in the days to follow I'd learn more about him. Like me, he had earned his private pilot license at sixteen. But he came from an aviation family; both his parents were also pilots. He had worked for US Airways for twenty-three years, with twenty thousand flight hours, and had risen to be a captain. But due to cutbacks in flights and planes, and the effect on the pilots' seniority list, he was now flying as a first officer. I have twenty-nine years under my belt, so these days, I'm among the most senior of pilots at my airline.

Jeff had been flying the Boeing 737 for eight years, and had just completed training to fly the Airbus. These seven flights over four days with me actually would be his first trip on the Airbus without an instructor. As Jeff put it, "It's my first trip without training wheels."

When I meet other pilots, I don't try to pigeonhole them. I figure I'll learn about them and their flying style in the cockpit. There's no need to rush to judgment. Still, my first impressions of Jeff were good ones.

From our initial moments together in the cockpit, for that flight to San Francisco, I found him to be conscientious and very well versed in everything about the Airbus. If he hadn't told me this was his first trip since being trained, I wouldn't have known.

Once pilots push back from the gate, and until we are above ten thousand feet in the air, cockpit crews aren't allowed to talk to each other about anything except the details of the flight. But after we were well on our way to San Francisco, Jeff and I were able to learn about each other. He told me he had three children, seventeen, fifteen, and twelve, and so we talked about our kids for a bit.

Somewhere over the snow-covered Rocky Mountains, I thought about that thrill I often get when I'm in the air, just taking in the majesty below, and the stars and planets around me, and appreciating all of it. It feels like we're floating through an invisible ocean of air, dotted with stars.

There's a poem I love, "Sea Fever," by John Masefield, which includes the line: "All I ask is a tall ship and a star to steer her by." I often think of that line when I see the planet Venus in the southwest corner of the sky as I head to the West Coast at certain times of the year. If I'm ever unable to access the global positioning system or use the compass in the cockpit, I know I'll be OK. I could just keep Venus in the left front corner of the windshield and we would reach California.

I mentioned to Jeff that I wished I could have my daughters take a flight with me in the cockpit of a commercial airliner, to see the pilot's-eye view of such scenes. In long-ago eras of aviation, that would have been possible. But in the wake of September 11, of course, restrictions on cockpit access were only increased. My girls will never see the skies through my eyes.

We also talked about our side jobs. Like a lot of pilots, Jeff also sees the need to supplement his income. He lives in Madison, Wisconsin, and has a business as a general contractor, building new homes.

Jeff said he'd Googled me before the trip because he was looking for my e-mail address. He wanted to share some scheduling information with me. Before the landing in the Hudson, of course, there wasn't much about me on the Internet. So the first thing he came upon was the Web site for my consulting business.

"I read all about your company," he said, and then he just grinned. "Man, I thought I was a good bullshitter, but you take the cake!"

I was intrigued that he had Googled me—I don't ever recall flying with another pilot who had—and I was also amused by how direct he was. "I consider myself a connoisseur of bullshit," he told me, "and you make that company of yours sound like it's this big operation. But then I read it more closely and I realized it's just you. You're the company. Good for you! I admire people who can take an acorn, and with a little bit of bullshit, make it into an oak."

I know my business isn't a Fortune 500 empire, but I'd argue a bit with his characterization. I really am passionate about safety issues, and about what the airline industry can teach the world. I'm proud of my work, and told Jeff that. Still, I got a kick out of his straight-shooting style. We had a good laugh about my fledgling consulting operation as we made our way to San Francisco.

Jeff was at the controls for a lot of the trip, and I was impressed by the ease with which he was handling things. We were aware, of course, that because he had fewer than a hundred hours on the Airbus, there were restrictions we had to follow. He couldn't land or take off where runways might be contaminated by snow or ice. And certain airports—because of high terrain or complicated takeoff or landing procedures—were off-limits to him. San Francisco was one of these airports, so I needed to land the plane there.

When we finally touched down on the runway at 8:35 P.M., I was back exactly where I'd started at seven-thirty that morning. But the good news was there were no flight delays; it was still early enough. There was time for me to get to my car in the airport parking lot, and drive fifty minutes northeast to Danville, so I could spend the night with Lorrie and the kids.

This was a bonus layover. Instead of being gone, as usual, for the entire four-day trip, I got to go home.

. . .

When I got into the house on that Monday night, it was nine forty-five and the girls were heading to bed. I didn't get to spend much time with them. But the next morning, I was able to drop them both off at school.

Kelly, now in eighth grade, had to be at her middle school by eight. I kissed her good-bye and told her I'd see her at the end of the week.

Then it was time to drive Kate to her high school. Actually, I was driven by Kate. She still had her driver's permit then, and was always looking to get experience, if not necessarily lessons. So she took the wheel and I got in the front passenger seat as a combination copilot and "check airman." That's the term for a pilot who is an instructor accompanying another pilot to assess his or her skills.

Being with Kate at the wheel of the family SUV was like being with Jeff on the Airbus. I was observing, admiring, and taking notes.

My take on Kate is that she's a good driver, though a bit overconfident. She's also not sure all the rules of the road apply to her, so I've tried to impress upon her the fact that driving laws prevent anarchy. In the airline industry, we'd say she's "selective about compliance." But overall, she's doing well. I'm pretty comfortable with her driving abilities, and told her so that morning. When she pulled up in front of her school, I kissed her and promised her I'd see her at the end of the week.

After I got back to the house, I made Lorrie a cup of tea and we had a pretty serious conversation. Because the Jiffy Lube franchisee had decided not to renew his lease six months earlier, and our commercial property—the land and the empty building—was still vacant, we were in serious financial straits. How long could we keep paying the mortgage without rent coming in? "Not much longer," I told Lorrie, and we discussed whether we'd need to sell our family home to solve our money problems. That would be a worst-case scenario, we agreed, and we had several other contingency plans for dealing with this before we'd have to sell. Still, it was a sobering and unresolved dilemma that would have to be tabled until my return later in the week. I needed to head back to the airport in San Francisco.

Before I left home, I made myself two sandwiches, one turkey and one peanut butter and jelly, and put them in a lunch bag along with a banana. This also has become part of my ritual. Until the last eight years or so, airlines provided meals for pilots and flight attendants on long flights. Economic cutbacks have ended that little perk.

On this day, because it was later in the morning, I was able to kiss Lorrie good-bye. An hour later I was at the airport again, preparing to pilot the A319 Airbus to Pittsburgh. Once Jeff and I got the plane into the air and on its way, those sandwiches and the banana served me well.

Much about flying has a hold on me. I still find it satisfying on many fronts—especially when I look out the cockpit window. I am grateful for all the adventures to be found at thirty thousand feet. But I've got to be honest: Eating PB&J while smelling the gourmet beef being distributed with wine in first class—that's a sure reminder that there are less-than-glamorous aspects of my job.

After we landed in Pittsburgh on that Tuesday night, I got in a van with Jeff and the flight attendants and we headed over to the La Quinta Inn & Suites near the airport.

We had to be flying again exactly ten hours later. This was close to what we call a "minimum night." Minimum rest for a crew overnighting between flights is nine hours and fifteen minutes. It sounds like enough time, but it's actually pretty tight. The clock starts ticking the minute the plane arrives and is blocked in at the gate. It continues until push-back of the next morning's flight. In between, we have to get out of the airplane, and to and from the hotel. We have to leave for the airport at least an hour, and sometimes ninety minutes, before the morning flight. Add in time for showering and getting something to eat, and our actual time sleeping is usually about six and a half hours.

Our flight that morning to LaGuardia Airport in New York left at 7:05. Because it was snowing, I handled the controls. We arrived at 8:34, got a new load of passengers, and were slated to head back to Pittsburgh at 9:15 A.M. Because of weather and traffic, we had a forty-five-minute delay on the ground at LaGuardia.

I still have my trip sheet from that week, and as always, I had scribbled notations alongside each flight. I keep track of all the actual flight times, to make sure I get paid properly. Pilots are paid per hour of flying, and "flying" is tallied from the moment you move away from the gate in one city to the moment you arrive at the gate in the next city.

Delays frustrate everybody—pilots, too, of course— but the fact is that we start getting paid when the plane has pushed back from the gate. If we sit on the tarmac for hours, we're getting paid. If we're waiting at the gate, we're not.

Anyway, we got back to Pittsburgh before noon, and because we had a long layover of twenty-two hours until the next leg of our trip, we were able to spend

Wednesday night farther from the airport, at the Hilton downtown. I went for a walk around Pittsburgh that afternoon by myself, bundled up in the snow, listening to my iPod. Jeff and I talked about having dinner together, but he had something to do, and so I was alone that night. The flight attendants were also on their own.

Because most US Airways flight crews are older now—no young blood has been hired for years—we're more tired and less social than we used to be. The wilder "Coffee, Tea, or Me" days are long over, and mostly predated my airline career. About a third to half of flight attendants and pilots these days are what those of us in the industry call "slam clickers"; they slam the doors to their hotel rooms and click the locks. They don't socialize and they spend their entire layovers in their rooms.

Granted, most of them aren't really slamming their doors. They say good night nicely and then disappear.

I understand that the constant travel is a grind, and that my colleagues are tired or don't want to go out on the town, wasting money. And I'm not a partyer by any stretch. But I decided a long time ago that if I was going to be gone from home sixteen or eighteen days a month—spending 60 percent of my time away from my family—I wasn't going to waste half my life sitting

in a hotel room watching cable TV. And so I try at least
to take a walk or go for a run. I'll visit a new restaurant,
even if I'm by myself. I try to have a life. If members
of the flight crew want to join me, I'm grateful for their
company. If not, I'm comfortable on my own.

On that Wednesday night, I called home and talked
to my daughters. I described my walk in the snow,
and asked them about what they were up to at school.
They are teens now, wrapped up in their own lives, so
they're not hugely engaged in hearing details of my day.
I'm always actively searching for ways to connect with
them, to keep things fresh.

The next morning, January 15, it was snowing,
and Jeff and I needed to take an Airbus A321 from
Pittsburgh down to Charlotte.

Because of the de-icing in Pittsburgh, we were thirty
minutes late arriving in Charlotte. And we switched
planes there, from an Airbus A321 to an A320. That
A320 was the plane that would take us to the Hudson.
The flight from Charlotte landed at LaGuardia just
after 2 P.M. It had been snowing in New York, but by
the time we arrived, the snow had stopped.

At LaGuardia, the gate agents started loading the
new passengers onto the plane. I got the flight plan
for the next leg—Flight 1549 from New York back to
Charlotte—and then ran to find something to eat. I

bought a tuna sandwich for eight dollars and change, and expected I'd get to eat it once we were at cruising altitude on our return to Charlotte.

Back at the gate, passengers had begun boarding, and I didn't get a chance to say anything to any of them. Some would later remark that I looked older with my gray hair, and they felt reassured that I was a veteran pilot. I just nodded and smiled at a few of them as I made my way back into the cockpit with my sandwich.

While the plane was being serviced, I checked the fuel load and the weather, and then went over the flight plan. As first officer, Jeff's job was to take a walk around the exterior of the plane, making an inspection. Nothing seemed out of the ordinary to either of us.

It was a full flight, 150 passengers, plus the crew— me, Jeff, Sheila, Donna, and Doreen. Just before we pushed back from the gate, Jeff and I remarked to each other that we had enjoyed flying together. This would be the final leg of our trip. I was planning to leave Charlotte at 5:50 P.M., flying home to San Francisco as a passenger, and Jeff was going to head back that evening to Wisconsin.

We pushed back from the gate at 3:03 P.M. Eastern standard time, and we joined the queue of airplanes waiting for our turn to take off.

In our ears, Jeff and I heard the constant chatter on the party line of the LaGuardia Tower Air Traffic Control frequency. We were listening in and watching as airplanes took off and landed on the two intersecting runways at one of the nation's busiest airports. As happens every day, it was a carefully choreographed ballet where everyone knew their parts exceedingly well.

At 3:20 P.M. and thirty-six seconds, the tower controller spoke to us: "Cactus fifteen forty-nine, LaGuardia, runway four position and hold, traffic will land three one." The tower controller was instructing us to taxi onto the active runway and hold in position to await clearance for takeoff. He was also advising us that we would see traffic landing on the intersecting runway 31. ("Cactus" is the radio call sign for US Airways flights. The airline chose it after we combined with the former America West Airlines. Though it was adopted to preserve the heritage of America West, some pilots and controllers would prefer that we had kept our old call sign, "USAir," to avoid confusion. Having a name that doesn't match the name painted on the side of an airplane can be confusing, particularly at foreign airports.)

At 3:20:40, as I was taxiing, Jeff responded to the controller: "Position and hold runway four for Cactus fifteen forty-nine."

We then sat on the runway for four minutes and fourteen seconds, listening to controllers and pilots trading concise esoteric exchanges such as "American three seventy-eight cleared to land three one, wind zero three zero, one zero, traffic will hold on four." This was the tower controller clearing American Flight 378 to land on runway 31, telling him the wind was from the northeast at ten knots, and advising him that Jeff and I were holding in position on runway 4.

At 3:24:54, from controller to me and Jeff: "Cactus fifteen forty-nine runway four, cleared for takeoff."

At 3:24:56, from me to controller: "Cactus fifteen forty-nine cleared for takeoff."

On the runway, shortly after we started rolling, I said, "Eighty," and Jeff answered, "Checked." That was the airspeed check. Our language was exactly by the book.

Then I said, "V1," an indication that I was monitoring the velocity of the airplane and that we had passed the point where we could abort our takeoff and still stop on the remaining portion of the runway. We were now obligated to continue the takeoff. A few seconds later, I said, "Rotate." That was my callout to Jeff that we had reached the speed at which he should pull back on the sidestick, causing the aircraft to lift off. We were airborne and it was very routine.

At 3:25:44, from the controller to me and Jeff: "Cactus fifteen forty-nine, contact New York departure, good day." We were being told that future communications for our flight were being handed off to the controller at New York Terminal Radar Approach Control, located on Long Island.

At 3:25:48, from me to the LaGuardia controller: "Good day."

To that point, my four-day trip had been completely unremarkable, and as with almost every other takeoff and landing I'd experienced in forty-two years as a pilot, I expected this flight to remain unremarkable.

We'd even made up a little time caused by the delays earlier in the day. So I was in a good mood. The Charlotte–San Francisco flight was still showing on time, and a middle seat was available. It looked like I'd make it home while Lorrie and the girls were still awake.

3

THOSE WHO CAME BEFORE ME

As human endeavors go, aviation is a very recent one. The Wright brothers first flew in 1903. That's just 106 years ago. I'm fifty-eight years old, and I've been flying for forty-two of those years. Aviation is so young that I've been involved in it for almost half of its history.

Through the efforts of many people in the past 106 years—their hard work, their practice, their engineering breakthroughs—aviation has quickly gone from its dangerous infancy to being so commonplace that there is little tolerance for any risk at all. We may have made it look too easy. People have forgotten what's at stake.

I'm not saying passengers shouldn't feel comfortable flying. It's just that it's easy to become complacent when our nation can sometimes go a year or two between major airline accidents involving fatalities. When things are going well, success can hide inefficiencies and deficiencies. And so it takes constant vigilance.

Long before I found myself in the cockpit of Flight 1549, I had closely studied other airline accidents. There is much to be learned from the experiences of pilots who were involved in the seminal accidents of recent decades. I have soberly paged through transcripts from cockpit voice recorders, with the last exchanges of pilots who didn't survive.

I studied these accidents partly because, in the early 1990s, I had joined a couple dozen other US Airways pilots to help develop an air-safety course looking at CRM—crew resource management. Before Flight 1549, my proudest professional contribution was my work in CRM. My fellow facilitators and I helped change the culture of our airline's pilot group by improving cockpit communication, leadership, and decision making. As First Officer Jeff Diercksmeier, my friend on the CRM team, said, "It was a time when a few people who really believed in what they were doing made a difference."

My interest in air safety goes back to my first flights as a teenager. I've always wanted to know how some pilots handled challenging situations and made the best decisions. These were men and women worth emulating.

And so I tried to understand, intimately, the full stories behind each of these pilot's actions. I'd ask myself: If I had been there, would I have been as successful?

A few years ago, I was invited to speak at an international conference in France focused on safety issues in a variety of industries. Given the comparatively ultrasafe record of commercial aviation, I was asked to appear on two panels to discuss how airline safety efforts might be transferable elsewhere. I talked about how other industries are recognizing that they can benefit by adopting some of our approaches.

This degree of safety requires tremendous commitment at every level of an organization and a constant diligence and vigilance to make it a reality.

Those of us who are pilots worry about the financial issues now weighing down airlines. Most passengers today select carriers based on price. If one airline's fare is five dollars less than a competitor's fare, the airline with the less costly ticket gets the booking. The net effect is that airlines are under intense pressure to lower their costs so they can offer competitive fares. This has cheapened the experience of flying; we've all seen the cutbacks in amenities offered in coach. But passengers don't see other ways in which the airlines are cutting back. For instance, some of the smaller regional airlines have lowered the minimum requirements for pilot recruitment, and they're paying some pilots $16,000 a year. Veteran pilots—those who have the experience that would help them in emergencies—won't take these jobs.

I have 19,700 flight hours now. Back when I had, say, 2,000 or 4,000 hours of experience, I knew a lot of things, but I did not yet possess the depth of understanding I have now. Since then, I've sharpened my skills and learned from many situations that tested and taught me. Regional airlines will now take someone with 200 hours of flying experience and make him or her a first officer. These new pilots may have exceptional training, and they may have a high degree of ability. But it takes time, hour after hour, to master the science and art of flying a commercial jet.

Another issue: Airlines used to have more large hangars in which their planes were repaired and maintained by their own mechanics. The mechanics would overhaul component parts, radios, brakes, engines. They knew the specific parts and systems in each aircraft in their fleet. Now many airlines have outsourced their maintenance and component work. Are these outside mechanics as experienced and knowledgeable about a particular aircraft? If a part is sent overseas to be overhauled, does it come back as reliable?

It's fair to say that when jobs are outsourced, and the work is done in a remote location, an airline has to work much harder to control the entire process, and to have the same level of confidence in the part or repair.

Every choice we in the airline industry make based solely on cost has ramifications and should be evaluated carefully. We have to constantly consider the unintended consequences for safety.

An airline accident is almost always the end result of a causal chain of events. If any one link was different, the outcome may have been different. Almost no accident was the result of just one problem. In most cases, one thing led to another, and then there was too much risk and a bad outcome. In aviation, we need to keep looking at the links in the chain.

Engine manufacturers know, for example, that their engines might someday encounter and ingest a flock of birds, causing severe damage. To learn what they're up against, the manufacturers use farm-raised birds to test their engines. These preslaughtered birds are fired into the spinning blades from pneumatic cannons—sacrificed in the name of research that might save human lives. Given the growing population of birds near many airports, this testing is crucial.

Birds certainly are entitled to their wide piece of the sky, but if we humans are to continue joining them there in ever larger numbers, we'll need to have a better understanding of the risks and remedies of bird strikes. In the wake of Flight 1549, investigators will likely consider whether an improvement in engine certification standards is needed.

Historically, safety advances in aviation often have been purchased with blood. It seems sometimes we've had to wait until the body count has risen high enough to create public awareness or political will. The worst air tragedies have led to the most important changes in design, training, regulations, or airline practices.

Airline disasters get massive media coverage, and the public's reaction in response to these tragedies has helped focus government and industry attention on safety issues.

People have incredibly high expectations for airline travel, and they should. But they don't always put the risks in perspective. Consider that more than thirty-seven thousand people died in auto accidents in the United States last year. That was about seven hundred a week, yet we never heard about most of those fatalities because they happened one or two at a time. Now imagine if seven hundred people were dying every week in airline accidents; the equivalent of a commercial jet crashing almost every day. The airports would be shut down and every airliner would be grounded.

In aviation, we should always aim for zero accidents. To come closer to accomplishing this, we must have the integrity to always do the right things, even if they cost more money. We have to build on all the hard work of the last 106 years, and not assume we can just rely on the progress made by previous generations. We need to keep renewing our investments in people, systems, and technologies to maintain the high level of safety we all deserve. It won't happen by itself. We have to choose to do this. This same prescription applies to many other industries and occupations.

Commercial aviation is one of several professions in which knowledge, skill, diligence, judgment, and experience are so important. With the lives of hundreds of passengers in our care, pilots know the stakes are high. That's why, long before Flight 1549, I read about and learned from the experiences of others. It matters.

When I arrived in the cockpit of Flight 1549, I would be aided by the courageous efforts of pilots who had come before me.

There were the two unheralded test pilots who, on September 20, 1944, risked their lives by landing their B-24 Liberator in Virginia's James River. This was a voluntary ditching, considered the first test on a full-size aircraft. As the plane hydroplaned for several hundred feet, which almost completely severed the bomber's nose section, engineers watched from a nearby boat, collecting data on how it fared. The pilots survived.

The following day, the *Daily Press* in Newport News had this headline: B-24 "DITCHED" TO EXPERIMENT ON STRUCTURES—JAMES RIVER TEST DESIGNED TO SAVE LIVES IN THE FUTURE.

By that day in 1944, the Allies had already ditched scores of bombers in World War II, often in the English Channel. Most filled with water and sank quickly; hundreds of crew members drowned. Better procedures for ditching were desperately needed.

As a recent *Daily Press* story explained, it took thirteen more years after that test in Virginia for a full report to be written on how best to attempt a water landing while piloting a distressed aircraft. That report called for landing gear to be retracted rather than extended. It described why an airplane should fly as slowly as possible, and why wing flaps should be down for impact. It also called for the nose to be up in most cases. These procedural guidelines remain in use today, and were in my head on Flight 1549.

As a student of history, I am awed when I read of the actions taken by these pilots in earlier eras. They didn't have all the data that now aids us in our decisions. They didn't have the benefit of all the additional decades of trial and error in aircraft design. They acted with the mental and physical tool kits available to them.

Perhaps the most famous water landing prior to Flight 1549 happened on October 15, 1956. It was Pan American Airways Flight 943, bound from Honolulu to San Francisco with twenty-five passengers. There were also forty-four cases of live canaries in the cargo hold.

In the middle of the Pacific, in the middle of the night, the Boeing 377 Stratocruiser lost two engines, and its remaining two engines were under strain, consuming large amounts of fuel.

Captain Richard Ogg, forty-two years old, knew he was too far into the trip to turn back to Hawaii. San Francisco was too far ahead. And so he opted for a water landing. He circled for several hours, burning off fuel and waiting for daylight, above a U.S. Coast Guard cutter that was in position to rescue passengers and crew.

Just before 8 A.M., the captain attempted his landing. The tail snapped off and the nose was shattered on impact, but all the passengers and crew were rescued. Captain Ogg went through the plane twice, making sure he didn't leave anyone behind. The plane took twenty-one minutes to sink below the surface of the Pacific.

The circumstances of Flight 943 were different from my experience on Flight 1549, mostly because Captain Ogg had hours to work on his plan and Jeff and I didn't even have minutes. Also, he was landing on the open ocean, not on a river. But I had long admired Captain Ogg's ability to safely land on water. I knew that not all pilots could have successfully equaled his effort.

After Flight 1549 hit the news, the *San Francisco Chronicle* contacted Captain Ogg's widow, Peggy, to ask her about the similarities between my landing in the Hudson and her husband's 1956 ditching in the Pacific. She spoke of her husband's sense of duty. He had told reporters at the time: "We had a certain job to do. We had to do it right or else."

When Captain Ogg was on his deathbed in 1991, his wife was sitting with him and noticed a faraway look on his face. She asked him what he was thinking about. He told her: "I was thinking of those poor canaries that drowned in the hold when I had to ditch the plane."

The first major airline accident I ever investigated personally was PSA Flight 1771, which crashed into hilly ranchland near Cayucos, California, on December 7, 1989. It was traveling from Los Angeles to San Francisco.

The specifics of the crash were haunting and disturbing. A former USAir ticket agent named David Burke, thirty-five years old, had been caught on a security videotape allegedly stealing sixty-nine dollars in in-flight cocktail receipts. He was fired, and tried unsuccessfully to get his job back. He then decided to buy a ticket on Flight 1771 because his supervisor was a passenger on it.

In that era before the September 11 attacks, those with airport IDs didn't necessarily have to go through security. So Burke was able to board the plane carrying a .44 Magnum revolver. Sometime after boarding, he wrote a note on an airsickness bag to his supervisor: "Hi Ray: I think it's sort of ironical that we ended up like this. I asked for some leniency for my family. Remember? Well, I got none and you'll get none."

The plane was at twenty-two thousand feet when the cockpit voice recorder picked up the sound of what appeared to be shots being fired in the cabin. Then a flight attendant was heard entering the cockpit. "We have a problem," she said. The captain answered: "What kind of problem?" Burke was then heard saying: "I'm the problem!"

The sounds of a struggle and gunshots followed. Investigators believed Burke shot the captain and first officer, and then himself, after which the plane went into a nosedive, probably because a pilot's body was slumped against the controls. The plane hit the ground at about seven hundred miles an hour and much of it disintegrated on impact. None of the forty-three people on board survived.

As an Air Line Pilots Association safety committee volunteer, I served as an investigator at the crash site as part of the "survival factors" working group, charged with trying to determine what the crew could have done to make that flight survivable. Of course, given the circumstances, there was almost nothing they could have done. The FBI quickly took over and turned the crash site into a crime scene. Over the days of searching, the handgun was recovered with six spent cartridges. So was the note on the airsickness bag, and Burke's identification badge, which he had used to avoid going through security.

When I got there, the crash site looked like an outdoor rock concert where everyone had left trash all over a hillside. There were hardly any big pieces of the plane besides landing gear forgings and engine cores. It was a very disturbing feeling being at the scene of a mass murder, knowing what had happened in the sky above us. The smell in the air was a mixture of jet fuel and death.

I had known one of the flight attendants on the plane, and it was horrifying to imagine what the crew and passengers went through. Working on this sort of investigation focuses your attention on how to prevent similar tragedies in the future. It renews your dedication to never let it happen again.

In the wake of Flight 1771, some groups of airline workers were subjected to security requirements similar to those set for passengers, better methods of employment verification were instituted, and federal law required employees to turn in their IDs after being terminated from airline jobs. But larger problems with security would still need to be addressed. Standing on that hillside in California, I couldn't have imagined the way cockpits would be breached on September 11, 2001.

In my role helping with accident investigations, I also was called upon to talk to passengers who survived crashes.

On February 1, 1991, there was a runway collision at Los Angeles International Airport between USAir Flight 1493 and SkyWest Airlines Flight 5569. It happened in part because the local air traffic controller cleared the USAir jet, a 737-3B7, to land while the SkyWest commuter plane, a Fairchild Metro III, was holding in position to take off on the same runway.

All ten people on the SkyWest plane died, and twenty-two passengers were killed on the 737. I was given the task of interviewing some of the sixty-seven survivors from the 737.

The NTSB gave us a long questionnaire, with questions such as: What announcements do you recall hearing? Did the emergency exit lights come on? Which exit did you use to escape? Did you help anyone else get out? Did anyone help you get out?

All of these questions were designed to help the airline industry learn from these events and improve the next outcome.

It was not especially pleasant work investigating accidents, but I was grateful for the opportunities to do so. When I talked to survivors, I listened carefully, trying to understand, and I filed away the details, in case I'd ever need to draw on them.

4

"MEASURE TWICE, CUT ONCE"

I grew up in a home where each of us had our own hammer.

When I think about the work ethic and the values that carried me through life, and through seven million miles as a pilot, I think at times about the hammer my dad gave me as a boy.

He had married my mom in 1948, bought a piece of farmland from her parents, and borrowed $3,000 to build a house on it. It was a very small ranch house, just one bedroom. But over the years that followed, my dad devoted himself to enlarging the homestead again and again. He built a series of additions with the help of three not-always-willing assistants: my mother, my sister, and me.

My parents were born in Denison, Texas, and my mom only lived in two homes her entire life, and they were within one mile of each other. The first was her childhood home, built around 1918 by my grandfather, Russell Hanna, who used materials he found right there on the property. He cleared the land of a great number of large stones, cut them with the help of a hired hand, and used them to build the house and other farm structures. From that home, my mom at age twenty-one moved just down the road to the little place she built with my dad. She'd live there, on Hanna Drive, for the rest of her life.

Certainly, my maternal grandfather could have named that gravel road First Avenue or Main Street or whatever. But the road led to his property, and so it bore his name. That's where I grew up, 11100 Hanna Drive, an ever-expanding house next to Lake Texoma, eleven miles outside of Denison.

My dad's father, who died before I was born, owned a planing mill—a final processing plant for lumber—and my paternal grandmother continued to be involved in the office operations after he was gone. It was right there in Denison, and when I was a young boy, I'd visit and play happily in the huge mounds of sawdust. The place was thick with the sounds of giant woodworking machinery and the wonderful smell of lumber. There was also a cool device on my grandmother's desk, a coil-springed gadget shaped like a human hand and made of stamped-out sheet metal. My grandmother stored envelopes and paperwork between the hand's fingers. Having grown up in that mill, my dad had a love and knowledge of woodworking, and of making things with his hands. By adulthood, he was a very able handyman.

That helps explain why, every few years when I was a kid, my dad would announce that it was time to enlarge the house. He and my mom would decide we needed a new bedroom or a larger living room. "Let's get to work," my dad would say, and we'd pull out the tools. He was a dentist, but he had taken drafting courses in high school. He had a big plywood drafting table he had made himself, and he'd sit there for hours with his T square and a pencil, drawing up plans. He was always reading *Popular Mechanics* and *Popular Science,* clipping articles about the latest home-building techniques.

The goal was to do everything ourselves, to learn what we didn't know and then have at it. My dad taught himself to do the carpentry, the electrical installations, even the roofing—and then he taught us. When we were doing the plumbing, my dad and I would heat the copper joints together, holding the solder, letting it melt from the tip of a soft wire. When we did electrical work, we knew we had to get it right: If we didn't, we risked electrocuting ourselves or burning down the house. None of this was easy, but it was satisfying on a lot of levels, and we were learning how to learn.

My father liked to use craftsmen's adages, such as "Measure twice, cut once." The first time I heard that particular phrase was after I had cut a piece of wood to go in the framing of one of our hallway walls. I cut it without paying close enough attention and it turned out to be too short.

"Go get another two-by-four," my dad told me, "and this time, measure more precisely. Then start over and measure everything again. Make sure you get a consistent answer. Then cut the board a little wide of the mark, just to give yourself an option. You can always make a board shorter. You can't make it longer."

I did as I was told, very carefully, and the board fit right where it belonged in the wall. My dad smiled at me. "Measure twice," he said. "Cut once. Remember that."

The four hammers in the house, one for each of us, got a huge workout. In the morning, before it got too hot, my dad would send us up on the roof to pound nails into the shingles. He never considered hiring a contractor or a roofing crew. For one thing, we didn't have extra money for that. And besides, as my dad saw it, this was a great family activity.

My sister, Mary, smiles at her memory of my dad driving us into nearby Sherman, where he had once come upon a certain house owned by a stranger. He loved that house. So when we were in grade school, he'd bring the whole family to sit in front of it while he sketched on a drawing pad, studying the parts of the structure that he liked. One day he'd sketch the roofline. A week later he'd come back and sketch the front steps. He wanted our house to look like that house, and he found his way by sketching the particulars.

My sister likes to say that watching my father expand our house showed her that anything is possible. "You can learn anything you want to learn," she says, "if you sit and figure things out logically, if you study something similar, if you keep working at it. You can start with a blank piece of paper and end up with a house."

This idea that "anything is possible" has been a bit of a mantra in my adult life, especially in my marriage. Lorrie reintroduced me to those words. And at the same time, my father's example remains there in the back of my mind, showing me the way.

That's not to say I always fully embraced my father's sense of the possibilities. On Saturdays, when my sister and I would have loved to sleep in, he'd wake us up at 7 A.M. so we could get an early start on whatever the latest expansion was. We'd work until lunchtime and then he'd suggest that we take a nap so we'd have the energy to get back to work later in the afternoon.

Even if we couldn't fall asleep, we pretended, so he wouldn't send us back to work right away. "Just keep your eyes closed," Mary would whisper to me. "He'll think we're still sleeping."

Though we dragged our feet at times, I did feel I had a stake in all of the construction work. I wanted to do a good job so all the additions would look right. Even in grade school and junior high, I felt committed to getting the masonry right, because I'd have to look at it every day. Also, I didn't want my friends to come over and notice that I lived in a place built by a bunch of amateurs.

The house was a source of pride, but I also felt a bit of embarrassment. Sometimes I'd brood, wishing we lived in a professionally built house like everyone else. I told myself that when I grew up, I'd live in a house where all the floors were completely level, where all the joints were square. To save money, my father also kept the heat low in the winter. I vowed to live in a house where it was never cold.

And yet, despite my mostly unvoiced complaints, I knew that working on the house was a special experience. Each time the place grew, I felt a sense of accomplishment. The house expansion was a tangible activity, not theoretical or intellectual. We saw the progress we made. We'd put in long days, especially in the summertime, but by nightfall, we could see that things were different from when we started in the morning. I liked that.

I've always liked seeing results. One chore I never minded doing as a boy was mowing the grass on our half-acre lot. When I was halfway through mowing, I knew how much I had left to go. When I was finished, I could tell I'd made a difference. The lawn looked neater. Flying for an airline offers equal satisfaction: We're halfway there. We've landed. We've completed our job.

My grandparents were all born between 1885 and 1893. All four attended college, which was especially remarkable for my grandmothers, given the times they lived in. My grandparents raised both of my parents with the belief that schooling was paramount, but that a lot also could be learned outside of formal education.

My father was born in 1917 and kept a journal when he was a teen that he later allowed me to read. The Depression became vivid to me as I paged through all of his journal entries. Money was always an issue, and he had a series of overlapping jobs in high school.

He'd balance his schoolwork with two paper routes and duties as a movie-theater usher.

My grandfather would sometimes run out of money at the end of the month, and he'd borrow money from my father. In his journal, my father chronicled his pluckiness, describing how he'd find ways to cope in hard times. When he had a little bit of money and could eat at the local diner, he'd order a bowl of chili and fill it with saltines and ketchup to make it a more substantial meal. It kept him from going hungry.

Reading my dad's diary, I got to better understand his worldview. It was a reminder of how much easier things were for my generation. I understood why my dad kept the heat turned down, and his kids hammering away at the house. Those with the Depression-era mentality never could quite shake it.

My dad ended up going to Baylor College of Dentistry in Dallas, graduated in June 1941, and decided to join the Navy. This was six months before Pearl Harbor was attacked.

He had always liked airplanes, and hoped to become a naval aviator. He even passed the rigorous physical exam. But then, at the last minute, he decided that since he had been trained in dentistry, perhaps he'd serve his country best as a dentist. It was a fateful decision. He entered the service with friends who did go on to become Navy pilots. They were killed in the fierce fighting early in the war. My father always assumed that if he had become an aviator, he would have been shot down with them.

He was stationed as a dental surgeon first in San Diego and then in Hawaii. He never was in combat, but plenty of men who saw the worst of it took their seats in his dental chair. Between 1941 and 1945, hundreds of those who'd been in battles told him their stories as they passed through Hawaii.

He took his work as a military dentist very seriously, and he learned things from the men who came through his dental office, especially the officers. When I was a boy, he would talk about the great obligations of a commander to look after every aspect of everyone's welfare who served under him. My dad made it clear to me how hard it would be for a commander to live with himself if, through lack of foresight or an error in judgment, he got someone hurt or killed.

When I was a boy, he impressed upon me that a commander's job is full of challenges, and his responsibilities are almost a sacred duty. I kept my father's words with me during my own military career, and after that, when I became an airline pilot, with hundreds of passengers in my care.

My dad left the service as a full commander, and after World War II, he opened a dental practice in

Denison. He loved talking to patients, and listening to what they had to say when his hands weren't in their mouths. But he wasn't much of a businessman. He had no ambition to run a large practice with a half-dozen associates, or to slave away for more than thirty-five or forty hours a week. Money didn't motivate him, and he never made too much or managed it particularly well. He didn't need a lot of material things, and figured we didn't either. Paying for my flying lessons was an indulgence, but he thought my time learning to fly with Mr. Cook gave me a sense of purpose and a path into the future. He was happy to find the money for that.

Unlike a lot of men of his generation, my dad thought of being with his family as his priority; work was secondary. I wouldn't say he was without ambition—after all, he built his own house—but he was content making less money if that meant he could spend more time with us.

It was almost as if he wasn't in dentistry to earn a living. A lot of the nuns from the local Catholic school were his patients. Sometimes they had the money to pay him, sometimes they didn't. He had other patients like that. Some people didn't get charged. Some didn't get charged much.

My father could also be a bit whimsical and impulsive. Or perhaps, as I'd later suspect, he was just looking for ways to brighten days when he was weighed down by darker moods. In any case, some mornings he'd wake up and say to my mother, "I don't feel like working today. Let's go to Dallas."

My mom would get on the phone and cancel all his patient appointments, then she'd call our school to say we wouldn't be coming in. My father figured my sister and I were smart kids; we could make up any missed schoolwork. And besides, he felt we could always learn something down in Dallas.

It was exciting. The whole family would drive the seventy-five miles listening to Top 40 songs on KLIF-AM on the car radio. When we got to Dallas, we'd see a movie and have an inexpensive dinner together.

We always stayed at the same little roadside one-story motel, a typical fifties-era row of rooms right off the freeway: the Como Motel. We'd swim in the small swimming pool in the middle of the parking lot. And we always ate at a Mexican restaurant called El Chico. Every meal, no matter what you ordered, came with rice and beans. I'd always get the cheese enchiladas, which I loved because of the diced onions inside.

El Chico had one large, open dining room with a high ceiling, and on the west wall was a huge mural of a Mayan—or maybe it was an Incan—outdoor scene. The focal point of the mural was a man with a native cloth around his waist and a bare upper torso. He was filling a jug with water, and I'd sit there eating my enchiladas and studying that guy in the mural. Every time we ate there, he was still filling that water jug.

We went to the same movie theater, the Inwood Theater, which had great air-conditioning at a time when it was a rarity in public places. That's where I saw two James Bond movies, *Dr. No* in 1962 when I was eleven, and *Goldfinger* in 1964 when I was thirteen.

Dallas was pretty cosmopolitan for us. It wasn't that large a city then, but it looked big to us, with its freeways and traffic and businesspeople walking around. John F. Kennedy was shot in 1963, and we may have driven by Dealey Plaza a few months after the assassination on the way somewhere. But we weren't gawkers. We didn't make a special trip to see it.

These spontaneous little getaways to Dallas were a continual reminder to me about my dad's appreciation of the distinctions between work and family. Family came first. Ahead of work and ahead of school. That's how my dad lived long before "work-life balance" became a popular catchphrase.

I have great memories of my family life closer to home, too. We had a small boat, and on weekends we'd take it out on Lake Texoma, which covered eighty-nine thousand acres. My mom was a terrific water-skier, and she could make it halfway across the lake, like the Energizer Bunny on skis.

We'd also take the boat out to one of the sandy islands in the middle of the lake. We'd camp out for the night, sleep in a tent, wake up in the morning and cook breakfast, and then we'd just cruise around in the boat. My dad often let me take the wheel—to captain the ship for the afternoon. I'd get badly sunburned, but it was all worth it.

One year, my dad bought a sailing magazine which had plans in it for a simple boat. He got out his woodworking tools, and we used the plans to build a dinghy out of plywood, with a bamboo pole for a mast and a bedsheet for a sail. I taught myself how to sail in that boat. I feel like my dad and I did almost all the projects I could have hoped for. Using our hands to create things, we shared a lot of great hours together.

People have asked if my dad is my hero. I never really thought of him in those terms. To me, he was just a great role model on a lot of fronts, from how he found his own ways to appreciate life, to the honorable way he conducted himself. He was always a perfect gentleman, a man who almost never raised his voice. I don't recall ever hearing him say a disparaging word about anyone.

Of course, looking back, there were sides of his personality that weren't easy to understand at the time. My parents never wanted us to see them fighting, or even having a frank discussion. They would go into the bedroom, close the door, and later come out presenting a united front. They went to great lengths to shield us from any bickering. So I never saw the messy details of how a couple might find agreement. As a young adult, I ended up with an unrealistic expectation that marriages were free of conflict.

There was something else about my father. He'd have days when he'd say he was in a "blue funk." He didn't fully explain himself, and outwardly, he seemed OK. But I now realize that he suffered from depression, probably for his entire life. In those days, when we thought of the word *depression,* we thought of the 1930s. The fact that being depressed could be a medical issue didn't occur to a lot of people. And so my father never got help, and just tried to cope with that "funk" on his own.

Sometimes that meant passing out the hammers and building an addition to the house. Sometimes it meant loading up the car and heading down to that roadside motel in Dallas. And sometimes it meant going alone into his room, where he'd deal with demons never discussed with the rest of us.

. . .

My mother was ten years younger than my dad, and especially at first, they had a very traditional marriage. She left college at age twenty-one to marry him, and later regretted not graduating. When I was a teen, she went back to school, majoring in education, and went on to get her master's degree. She taught kindergarten at first, and then spent most of her career as a first-grade teacher at Sam Houston Elementary School in Denison.

It was a great kick to be my mom's son in Denison. People tend to love their first-grade teachers, and my mother was especially kind and nurturing with kids. She was absolutely beloved in town. It's not a stretch to say she was something of a minor local celebrity.

My mother was a terrific pianist, too, and I just loved listening to her play Chopin. When I was in grade school, I'd always say to her, "Will you play more Chopin?" I'm not sure a lot of kids today, plugged into their iPods and cell phones, are calling out to their mothers for more Chopin. But my mother helped instill in me an appreciation of classical music. She was my favorite performer.

I always like to say that my mother gave me three important things: a lifelong love of reading, learning, and music. These are three very special gifts.

I also saw in my mother a commitment to service. She was a leader in a local chapter of the woman's group PEO (Philanthropic Educational Organization). Founded in 1869 in Iowa, its mission was to promote educational opportunities for women. In my mother's day, there were plenty of people who didn't think much of the idea of women going to college, and PEO's platform was somewhat controversial in some circles. And so my mom was very secretive about this PEO "sisterhood." She wouldn't tell me what they stood for, what they did, what happened at their meetings, or who attended. There was a desire by these women to be quiet about their work. Looking back, I salute them for the work they did to encourage young women to fulfill their potential, but I realize theirs was a form of feminism that hadn't yet found its full-throated voice.

My mother was an advocate for children, too. She believed that young kids could handle more responsibility than adults might imagine. She saw this in her first graders, but she felt it long before she was a teacher.

From the time I was very young, she and my dad impressed upon me the importance of looking after my sister, who was just twenty-one months younger. My father had the traditional sense that men should take care of women. And so he anointed me a kind of "second dad." But my mother just thought children can rise to the responsibilities they're given.

"When we're not around, we're counting on you," my mom would tell me. My dad would say, "You're in charge."

I wasn't always the perfect older brother. When I was five and Mary was three, I once took her out to play in the gravel on Hanna Drive. Some of the stones were smaller than a pea, and I thought it would be fun to feed these tiny stones to my sister. My mom caught me and told me a five-year-old should know better than that. Maybe I did know better, but at that age, feeding gravel to your kid sister doesn't necessarily seem like a bad way to pass the time.

My sister now says that I was a pretty good big brother most of the time. She thinks that looking after her helped me develop the sense of responsibility that has carried me through life, and into my career as a pilot. A couple of times as a teen, she went out with guys who were too forward, or who weren't completely respectful. I took it upon myself to go talk to them and set them straight. My sister feels that even when the two of us were arguing, I was protective and committed to keeping her safe.

We weren't a hugely demonstrative family when it came to showing affection. But we were there for one another, and we felt a sure sense of duty. We also had faith in one another. My mother knew my capabilities and encouraged me to have confidence in them. That's why she was comfortable flying as my passenger when I was a teenager. She knew that I knew I could do it.

My sister also was never afraid to fly with me. "Maybe it's the invincibility of youth," she now tells me, "and I just figured nothing could happen to me. But I think the main reason I had no fear is because I had an innate confidence in you. I knew you'd protect me."

I was very sure of myself and directed as a kid in the 1960s. I expected to serve in the military and then be a commercial pilot. Looking back, I think I was a very earnest, serious boy still struggling to figure out where I fit in the world.

In one eighth-grade school essay, titled "The Way I Am," I wrote: "I have good habits as well as bad ones. Being polite is one of my good points. My parents have taught me the manners I should know. I think my table manners are what they should be.

"I have bad habits, too. I am not very patient sometimes with other people. I would like to do everything exactly right, and I would like others to do the same. I should realize that everyone is not perfect.

"I know many people who have better personalities than I do, but I am doing the best I can."

My teacher wrote at the end of the essay: "You are doing fine." That's the way things went in those days. Teachers and parents didn't spend a lot of time stroking kids, telling them they were special. Back then, "you are doing fine" was what passed for a compliment.

I see my adult self in that essay. I remain regimented, demanding of myself and others—a perfectionist—though I think that has made me a better pilot.

In another essay, celebrating my family, I wrote about my sister "of whom I am proud, despite her behavior at times." I wrote of how fortunate I felt to be my mother's son: "She cares for me day and night." As for my father: "He guides me and teaches me and makes me wiser and more able to profit from my mistakes."

In the end, it didn't matter that some of the floors in our house were slanted, or that my dad wasn't paying attention to making money. I was supremely lucky to grow up on Hanna Drive, to know where every nail was, and to be nurtured and taught by two people who got so many things right.

5

THE GIFT OF GIRLS

I have seen breathtaking sunrises and sunsets from the highest altitudes. I have seen the brightest stars and planets from what feels like a front-row seat. But there are things I haven't seen—things that happened down on the ground while I was up in the air, earning a living and appreciating the view.

Being away from home so much, I've missed milestones in my daughters' lives. Many pilots can recite a litany of missed moments. Our children don't wait for us before they take their first steps, say their first words, or need a visit from the tooth fairy. And it's not just early-childhood rites of passage that we're sorry to miss. We also miss nuanced changes in our children's lives as they get older.

Just before Christmas last year, I was off for a few days, and Lorrie and I took our daughters, Kate and

Kelly, on a skiing vacation at Lake Tahoe. It was so nice to have this extended time with the girls when they weren't rushing off to school and I wasn't hours away from returning to the airport. It was just a perfect, relaxed vacation.

Tahoe has always held a special place in our hearts. When we take Interstate 80 and cross over the Donner Summit, a part of us feels like we've come home. There's the smell of pine in the air. The sky is clear and crisp. It's just invigorating.

We always try to stay at Northstar, the resort where both Kate and Kelly learned to ski when they were three years old. The resort resembles a European village with cobblestone walkways, and the family programs there are great. We have many wonderful family memories of visits there.

On that particular trip, the first big snowstorm of the season had ended the day before, and the trees were still heavy with fresh snow. Decorated for the holidays, Northstar was covered with little twinkling white lights in the trees. It had a real magical, fairy-tale feel. The lights, the snow, the European village.

Late one afternoon, we had just parked the car, and we decided to do some window-shopping before heading to dinner. It was very cold out, and we were all dressed in heavy jackets, gloves, and hats. We were walking into this valley of buildings, on this cobble-stone walkway, when I noticed that the girls, twenty feet ahead of us, were arm in arm and skipping along the sidewalk, Kelly's head on Kate's shoulder. I was so happy to see this, to realize that they had come to a place, here in their early teens, where they could publicly show physical affection for each other. Siblings, of course, are sometimes at odds, and here they were expressing so effortlessly what they meant to each other.

I pointed them out to Lorrie. "Take a look at that," I said. I thought I was noticing something very special and new.

Lorrie took my arm and smiled. "They've been doing that for five or six months now," she said. "It's just that you've missed it."

She said she had frequently seen them walking in the mall, holding hands. She said it was happening very easily and naturally, and she had loved watching it.

I had never fully noticed this. Not until that afternoon. And I felt sadness at the realization of how much of their daily lives I had missed—their activities, their interactions. How could I have missed witnessing these acts of love between my daughters for all these months? Lorrie looked at me sympathetically and saw a sense of loss and remorse in my eyes.

I put my hand over my heart. It's a gesture I sometimes fall back on when the girls do something endearing, or that I feel grateful about. It's a sign between me and Lorrie, a reminder of how lucky we feel about our girls.

I know why it hit me so hard. This was almost like a dream come true. When the girls were very young, one wish Lorrie and I had for them was that they'd be close when they were older. Seeing them together like this was a wonderful realization; I felt like maybe we had done something right. But it was also a painful reminder to me that I am so often not present in my children's lives.

Lorrie says this was one of those "pilot moments"— a pilot comes home and notices a change in his home or family—and seeing my mixed emotions was emotional for her, too.

I took Lorrie's hand, and a few seconds later we made a right-hand turn and came upon a large plaza in the village. Laid out in front of us were twinkling lights. Holiday music was playing and people were ice-skating and roasting marshmallows. There was a large outdoor fire pit. I held tight to Lorrie's hand and enjoyed all of it.

When I go over that day in my mind, I think of the girls, but I also think about Lorrie. I know what a loving mother she is. Yes, I've tried my best to instill values in the girls, to help them find more reasons to care about each other. But Lorrie is on the front line, nurturing them, setting an example, being there for them day and night when I am far away. I marvel at how she has created such a wonderful home life for our family.

I am fortunate to be her husband, and to have her as the mother of my children.

July 6, 1936, is a red-letter day for me, and not just because it's the day federal air traffic control began operation under the Bureau of Air Commerce.

Yes, I'm taken with the history, but that day stands out for me on a more personal level. Fifty years later, on July 6, 1986, a fiftieth-anniversary celebration was held at the Oakland Air Route Traffic Control Center in Fremont, California. Organizers invited the public to tour the facility, to see where controllers direct the flow of air traffic over Northern California. Pacific Southwest Airlines agreed to send over a pilot and a flight attendant to talk to the guests, and I was asked to be the pilot on hand.

I had flown a red-eye the night before, as a first officer, so I'd been up a lot of hours and was pretty beat. But I was more than happy to explain how pilots interact with air traffic control.

The flight attendant who had been selected to join me got sick and couldn't come. So PSA sent over a vivacious twenty-seven-year-old from its marketing department, a young woman I had never met before. She told me her name was Lorrie Henry, and I introduced myself.

"Hi, I'm Sully Sullenberger."

I have an uncommon name that she must not have heard clearly, and she never asked me to repeat myself. So that entire day, she didn't know how to address me. She just knew I had a lot of *S*s and *L*s in my name.

Lorrie will tell you it wasn't love at first sight. Despite my pilot's uniform, I looked tired, and she noticed my eyes were bloodshot and I wasn't freshly shaved. And she kept thinking: What's this guy's name again?

At the time, Lorrie had sworn off dating. She'd had a few relationships she considered unhealthy, and had told herself she was taking a break from men. I was thirty-five years old, had been in a short, childless marriage, and I wasn't exactly looking for long-term love either. But I was taken with Lorrie. She was attractive—tall and elegant, with a great smile—and she seemed smart, too. She was very engaging with all the passersby. Pretty quickly after meeting her, I knew I wanted to ask her out.

For about four hours, we stood next to each other greeting the public beside a large model of a PSA aircraft, the BAe-146. A lot of people who came by wanted to share tales of their most memorable PSA flights.

Lorrie wasn't at all flirtatious toward me, and I also remained professional toward her. But I was waiting for my moment. As the event wound down, I said to her, "Why don't we go get a drink?"

"There's a commissary down the hall," she told me. "If you're looking for a vending machine, you can find one there."

She wasn't getting it, but I wasn't giving up just yet. "I meant a cocktail," I said. "In a bar."

She looked at me, this weary pilot with a lot of *S*s and *L*s in his name and a confusing opening line, and I suppose a part of her took pity on me. She agreed to accompany me to a nearby Bennigan's. We had that drink, talked for a bit, and as she'd later admit, there wasn't any wild attraction on her part. She assumed she'd never see me again. But I was interested. I asked for her phone number and she gave me her PSA business card, which had only the 800 number on it for the airline's marketing department.

I tried to be clever. "You must be in great demand," I said, "if you have your own eight-hundred number."

She resisted rolling her eyes at me and just smiled, and then she gave me her local phone number. I gave her my card and she finally saw my name spelled out. We made a date for a couple of days later.

By the time Lorrie got home, however, she had decided she wasn't ready to date anyone, and in any case, she wasn't really up for dating me. She called me and left a message on my answering machine that she had to work the night of our date.

Listening to her message, I clearly sensed her lack of interest, and I figured that was that. But days later, Lorrie told a close friend that she had decided not to go out with me. Her friend told her, "No man is going to find you if you're sitting home on the couch."

Lorrie argued that the couch was just fine for her. She wasn't looking for a man, anyway. Still, her friend's words stayed with her, and a week later, she was surprised to find herself calling me.

When we spoke she admitted that she had been less than truthful when she canceled on me, and that she was nervous making the call. She said she'd like to accept a date with me if I was still interested. Of course, I was.

We lived fifty-five miles apart, but we ended up seeing each other for dinner three Friday nights in a row. After the second dinner, I walked her to her car, leaned toward her, and kissed her. Lorrie thought I was being forward. The way she tells the story now, she was "taken aback a little bit." But I kissed her for a reason. I wanted her to know that I wanted to kiss her, and that I found her attractive. I'm glad I kissed her. I'd do it again. (In fact, I have.)

That kiss was a turning point, and she began warming up to me, too. For more than a year, we went back and forth between her home in Pleasant Hill and mine in Belmont. Eventually it just felt right to move in together. In early 1988, we settled into my place.

I'll never forget coming home to Lorrie for the first time after being away on a four-day trip. The house was glowing. She had music on, the food on the stove smelled wonderful, and the house was warm and inviting. "If I had known it would be like this," I told her, "I'd have insisted we move in together sooner."

Marriage was the obvious next step, and on the morning of our wedding, June 17, 1989, I wrote Lorrie a letter: "I can't wait to marry you. I want you and need you and love you with all my heart."

I meant every word of that, but it's hard for a groom on his wedding day to fully understand all the challenges of marriage. Lorrie and I would have to learn to face a lot of obstacles together. There were adventures ahead that we never could have predicted.

. . .

Lorrie provides a lot of the color in our lives. She's intuitive, emotional, creative, more at ease with people, and more outgoing. In certain ways, she's more innately optimistic than I am. It can take a lot to get me to smile, but you'll often find Lorrie walking around with a smile on her face for no particular reason. Before Flight 1549 made me recognizable, we'd go to parties and everyone would remember Lorrie. As for me, couples would drive home saying to each other, "I think he said he was an airline pilot."

I'm analytical, methodical, more of a scientist. I am able to fix things. I'm optimistic if I've reviewed the information and decided that I can make something work. Otherwise, I'm pretty much a realist. Together, Lorrie and I like to say, we become one whole person. So in a lot of respects, we're a good fit.

Of course, our differences also get in the way. "When you're the emotional one, you want your spouse to emote more," Lorrie says. I do try, but I'm not always good at it. She wants to have detailed discussions about our relationship and our family dynamics. I'm more specific. What are the issues? What steps can I take to correct a problem?

I've asked Lorrie: "If things are going OK, why do we need to talk about them so much?"

I can feel close to Lorrie by touching her hand or giving her a hug. I'm nonverbal. She says it takes more effort than that to have a real relationship—and that means conversation.

I try. But sometimes, by the end of the day, you can feel you've said everything you've wanted to say. I've had to learn that it's important to save something for Lorrie—an anecdote, something I've read, something funny that happened on a trip. Lorrie has discovered that I become a better talker when she gets me out of the house and into the fresh air. When we take a hike or walk together, she says, it's easier to engage me in conversation.

We also try to have regular date nights, and we make a point of dressing up, rather than wearing casual clothes all the time. It's a way of showing respect; we're not taking each other for granted.

Lorrie likes me to make the reservations once in a while so I'm not always leaving it to her to be the social secretary. And when we go out, she wants to have a real dinner conversation.

"Sully is a man of few words," Lorrie tells her friends. "So I tell him to save up his words for date night."

Lorrie says that part of what makes me a good pilot is my attention to detail. She has told me: "Sully, you expect a lot from yourself and those around you. You're in control. That helps you as a pilot. But those aren't always good husband qualities. Sometimes I need a companion who is more forgiving and less of a perfectionist."

I know I can be exasperating to Lorrie. "Sully," she has said more than once, "life is not a checklist!"

I understand her frustration, but I don't see myself that way. I'm organized. I'm not a robot.

She says that when we go on vacation, I choreograph things with military precision, from loading the trunk to the time of departure. "That makes sense if you're flying a hundred fifty passengers to some vacation destination," she tells me. "But if you're just packing our suitcases into the car for a family getaway, it's not necessary."

My response to her: "That's confirmation bias. You find things that confirm your point of view, and you ignore evidence to the contrary."

In my heart, of course, I know she has a valid point.

In some important ways, my profession as a pilot is easier for me than relationships are. I can control an airplane and make it do what I want it to do. I can learn all of its component systems and understand how they work in every circumstance. Piloting is well defined, with a process that is predictable and understandable to me. Relationships, on the other hand, are more ambiguous. There's a good deal of nuance, and it's not always obvious what the right answer is.

In the twenty years of our marriage, we've had our share of bumps in the road. At certain points, one of us would be working harder at the relationship than the other, and then it would flip-flop. We weren't always equally committed to addressing issues. That has been an impediment at times.

Lorrie describes herself as "the voice raiser, the emotional one." I'm easily frustrated, often tired from traveling. And the fact that I'm always packing up to leave doesn't help. Marriage counselors advise couples not to go to bed angry. It's also not a good idea to fly across the country angry, leaving an unhappy spouse at home.

"For me, absence does not make the heart grow fonder," Lorrie says. She stopped working at PSA a long time ago, and has spent most of her energy since then as an at-home mom. She would love to have a husband who comes home every evening. "We could have a glass of wine, eat dinner together, chat about our day," she says. "And I don't even need the wine or the meal. I just want the husband in the room with me." She and I have nice phone conversations when I'm on the road. "It's not the same as having you here," she tells me.

In some ways, it was worse when the kids were younger, because back then Lorrie wanted my hands-on help. For a while we had two in diapers and in car seats, and she felt overwhelmed when I left on a long trip. Sometimes, she'd be in tears as we said our good-byes. In her PSA days, she had once gotten to sit in a flight simulator. "I know the flap settings," she'd tell me. "I'll get the plane off the ground. You stay home with two crying babies for four days." She was joking, but . . .

Now that the kids are older, she says that when I return home after a four- or five-day absence, my re-entry to family life isn't always smooth. I'm jet-lagged, I'm out of the loop of family activities. I've missed a lot. Lorrie says it sometimes takes me a day and a half before I can give something back to the relationship. I'm in the house, but I'm not able to jump back into our normal routine with the same vigor. Sometimes I'm just feeling spent, and not eager to attend to household chores.

I do see myself at times as somewhat of an outsider in my own family. But I love that the girls connect so well with Lorrie, and I understand why my bonds with them are not as effortless. I get it: I'm more formal, I'm male, I'm older, I'm gone a lot.

Parents build up a bank account of interactions and memories with their children. Lorrie has had a lot more moments with the kids than I have, so her bank balance with the girls is higher than mine. Certainly, there's a lot of love between me and the girls, but I know I have handicaps that I have to work to overcome.

My time away is a challenge. But Lorrie and I have been through great challenges together, and we have spent twenty years working through them. We work hard to find the right balance. We have both learned a lot about ourselves and each other and about what it takes to make a relationship work and to make it rewarding. We have both grown. By working on this together for each other and for our girls, we have become better people. We have invested in ourselves.

How did my personal life, apart from my aviation experiences, prepare me for that journey to the Hudson? I think that these challenges Lorrie and I faced together made me better able to accept the cards I've been dealt—and to play them with all the resources at my disposal. Early in our marriage, Lorrie and I were dealt the challenge of infertility.

A year or so after we got married, Lorrie and I began planning to have a family. We spent a year trying to conceive, without success, and then went to a fertility specialist. For six months, Lorrie took Clomid to induce ovulation. Like many women on that drug, she gained weight, and that was troubling for her. She'd been in good shape before starting on the medication, and here, for reasons beyond her control, she just kept getting heavier. She put on thirty-five pounds.

One day she and I were in the car and she turned to me and said, "You never make a comment about how I look or about my weight." My reply came naturally to me—I just said what I felt—but it meant a lot to Lorrie. I told her: "You don't get it, do you? I love you for what's on the inside."

"That's what every woman wants to hear," she said, and she meant it.

Sometimes I get things right.

We kept trying to conceive, but I was off on trips a lot, which made it hard for Lorrie and me to connect at the appropriate moment. A couple of times, she flew to the city where I was staying on a layover so we wouldn't "waste" a thirty-day cycle. It wasn't exactly romantic. We were focused and a bit tense. We were on a mission.

The Clomid didn't work, so eventually we turned to in vitro fertilization. The cost was $15,000—not covered by insurance—and we were told the success rate was about 15 percent. Lorrie needed to endure shots at 2 A.M. and 2 P.M., and when I was home, I'd give them to her. When I wasn't home, she gave them to herself.

These were not easy times for Lorrie. "I feel like my body has betrayed me," she'd say. "My body won't do the one thing it was designed to do, the one thing that separates one gender from the other." We'd been raising guide dogs for the blind, and a couple of the dogs were pregnant at the time. "It seems like everyone and every animal I meet is pregnant," Lorrie would tell me. "Everyone except me." I knew she felt deeply wounded, but I didn't fully know how to help her.

I was the one who had to tell Lorrie that the in vitro effort hadn't worked. She took one look at me and she knew. I had what she later described as a completely flat expression on my face.

I felt devastated for myself, but even more so for Lorrie. All I could say to her was: "Honey, I'm so sorry." We hugged each other and she cried for a while. I tried to be stoic for her, but I was hurting, too.

We went back to the doctor, who told us we were both still relatively young—I was thirty-nine and Lorrie was thirty-one—and we should consider trying again.

Lorrie had gotten to know another woman who was a patient at the clinic, and on the day Lorrie learned she wasn't pregnant, that woman was thrilled to learn she was. But then, a few days later, the woman was told that actually her pregnancy hadn't taken. It was possibly more devastating to have such high hopes dashed. When Lorrie heard this news, she decided she'd had enough.

"What's our main goal?" she asked me, and then she answered. "Our goal isn't for me to be pregnant. Our goal is to have a family. And there are other ways we can do that."

Before she met me, Lorrie had been a longtime Big Brothers Big Sisters volunteer. She saw that as both a duty and a labor of love. She began mentoring her "little sister" when she was twenty-six and the girl was five. Now Lorrie is fifty and her little sister, Sara Diskin, is twenty-nine, and they're still close. And so when Lorrie was unable to get pregnant, she was able to frame our predicament very clearly. "I've known for a long time," she told me, "that the beauty of a relationship is not biology. I'm ready to move on."

And so we decided we'd adopt.

Trying to adopt a baby was also an arduous journey—a long, difficult, emotional, expensive roller coaster—and we learned a lot about ourselves in the process.

Lorrie vowed to approach the adoption search as a full-time job. It took effort to educate ourselves about a process that was not well defined. There were many avenues. Which ones would pay off? Lorrie tried to have a business plan, but adoptions don't always proceed logically.

The fortunes of adoptive parents vary according to the wishes of birth parents. Their names are buried deep on waiting lists, while their files get dissected at agencies by people who don't really know them. There's no clear order to the process.

Lorrie was very emotional through all of it, and my attempts at a workmanlike approach didn't always help. "You don't know how to console me," she told me at one point. "It's outside your parameters. You're unable to feel things the way I feel them."

Lorrie struggled with all the paperwork we had to file, and the fact that we had to "qualify" to be adoptive parents. It was hard for her. Throughout her infertility treatments, she was poked and prodded. She had surrendered her body in an effort to find her way to parenthood. She had shown her commitment. Now she was being asked to find friends who'd vouch for whether or not she could handle being a parent. It felt almost like an insult.

Lorrie and I handled all the paperwork very differently. One day we exchanged our answers to a set of questions. I had to tell Lorrie: "You're overthinking this. Just answer the simple question with a direct answer." She was grateful when I told her that. It allowed her to temper some of her anxiety about the process. She didn't owe them her life story. She owed them basic answers to their questions.

We met with several sets of birth parents over the months that followed, hoping they'd select us. That was a hard process, too. Lorrie would often be excited after a meeting, certain that we'd get the nod. I tried to be logical and analytical. "Yes, that birth mother said a lot of nice things about us," I'd tell Lorrie, "but think about what she didn't say." Lorrie said I was raining on her parade, but I felt we had to look at everything realistically or we'd set ourselves up for wave upon wave of disappointments.

We met with a variety of birth parents during our search. And then, on December 1, 1992, we flew down to San Diego to meet a woman who was seven months pregnant. The birth father was there, too.

The couple asked us about our lives, our dreams for the child we hoped to someday raise, my schedule as a pilot, everything. They were honest and clear-eyed as we spoke, and so were we. Not long after that, we got word: They had selected us to be the adoptive parents.

At 2 A.M. on January 19, 1993, we got a call that the birth mother was in the delivery room, and we should prepare to fly down to San Diego to pick up our new baby. Lorrie was too excited to sleep. As for me, the realist, I knew that I'd be a better father in the morning if I got some sleep. So I went back to bed. Lorrie couldn't believe how I could sleep at a time like this. She stayed up, sitting by the phone, waiting.

Kate was born at 4 A.M., and we flew to San Diego just after sunrise. We brought a car seat with us because we'd need it in the rental car once we picked up the baby. Lorrie and I felt a little self-conscious walking through the airport with that empty car seat. Were people looking at us, wondering where our baby was?

When we arrived at the hospital, we went straight to the nursery and saw Kate for the first time; it was an overwhelming moment. I fell in love with her the second I saw her.

Later, a nurse was holding Kate. "Would the mother like to hold the baby?" the nurse asked. The birth mother pointed to Lorrie and said, "She's the mother." Lorrie was handed Kate.

Eventually, Lorrie had to use the bathroom, and while she was gone, Kate needed to have her diaper changed. I was proud to be the first of us to get to do that.

Early that afternoon, hospital staffers told us we were free to take Kate and go. Lorrie wanted to say good-bye to the birth mother. "What can you say to a woman who has given you this kind of gift?" she wondered. "I don't think there are any words."

Both of us considered the birth parents to be incredibly courageous people. They knew that for whatever reason—their age, circumstances, finances—they couldn't raise their child. And so they had made a very hard yet loving choice. They had turned their wrenching dilemma into a gift.

Lorrie left the baby with me in the nursery—she thought it would be too hard for the birth mother to see Kate one last time—and she went into the birth mother's hospital room. As she offered a simple thank-you, she saw a single tear running down the birth mother's face.

"Just be good to her," the birth mother said.

It was an overwhelming moment for both of them.

Hospital protocol requires new mothers to leave the hospital in a wheelchair. Lorrie tried to explain that she hadn't given birth and didn't need a wheelchair, but the aide with the wheelchair insisted on accompanying us out the front door. And so we walked, holding Kate, as the empty wheelchair was pushed beside us. It was ridiculous and surreal, but it was an amazingly happy moment, too.

In the parking lot, it almost felt as if we had stolen Kate. We looked over our shoulders, wondering if someone would be coming back to get her. We ended up putting her in our car seat, driving a mile from the hospital, and pulling over to the curb.

We looked at each other. We looked at Kate, who looked up at us. I wasn't crying, but it was as emotional a moment as I've ever had in my life. I was a father.

Just fourteen hours after being born, Kate was on her first airplane ride, heading back with us to Northern California. As an aviator, I was certainly happy to get her into the air that quickly.

Two years later, another birth mother looked through thirty-six bios in a book of potential adoptive parents, and after meeting Lorrie and me, agreed to make us parents for the second time. On January 6, 1995, when the call came that the birth mother had gone into labor with Kelly, I was in Pittsburgh, receiving simulator training on the MD-80. I cut short my training and made plans to return home as soon as possible, which was the following morning.

Lorrie, meanwhile, headed to the hospital. For the birth mother, it was a very long labor, and Lorrie stayed up for twenty-four hours straight, just waiting. Unlike when Kate was born, this time Lorrie was in the delivery room, and the whole day had a cinematic feel to it. There was a huge storm outside, with rain coming down in buckets and a howling wind. Then, when Kelly was finally crowning, a nurse gasped and said, "Oh my gosh!"

Lorrie was taken aback. "What, what, what?" she said, her heart pounding.

The nurse answered, "We've got a redhead!"

As soon as Kelly arrived, just after 10 A.M., the doctor handed her to Lorrie, which was an overwhelming moment for her. The rain. The thunder. This new beautiful baby. And I missed it all. While Lorrie was cuddling Kelly in the first seconds of her life, I was above the clouds somewhere over Denver.

I made it to the hospital that afternoon, and seeing Kelly for the first time was another moment of instant love and gratitude. And the most amazing thing was how much Kelly looked like me when I was a baby: the shape of our heads, our eyes, our Irish coloring. I was strawberry blond as a boy. We'd later mount baby photos of me and Kelly side by side in a frame, and it was hard to tell us apart. It's interesting how that goes in an adoption sometimes. Lorrie likes to say that we are blessed to have children who resemble us. It's not that we need the girls to look like us, but it's nice that they do. And over the years, it meant that if we opted not to voluntarily tell various people about the adoptions right away, we didn't have to.

Kelly's adoption was more complicated than Kate's. There are a lot of factors that can slow down the paperwork—or even make it fall through. It's hard for birth mothers to make their decisions final. They often have family pressures to consider.

Lorrie and I had to deal with some of these issues, and we struggled with the uncertainty. We passed the hours at a restaurant called Taxi's, which was near the hospital. We ate lunch and dinner there while we anxiously waited for the paperwork to come through. We were deathly afraid, with time passing, that some bureaucratic snafu could lead other issues to unravel, and keep the adoption from being finalized. At one point, I had a very forceful conversation with the hospital administrator, telling him that the hospital had to get its act together. I was pretty worked up and assertive, but it was necessary to break the logjam.

On the day we brought Kelly home, we had her in a car seat in the back of our car. Two-year-old Kate came out of the house and stared quizzically at this baby. She thought Kelly was a new doll she was getting as a present. She'd soon know better.

Out there at the car, Lorrie and I looked at each other and I said what I was thinking: "We're a real family now."

As we get deeper into our marriage, Lorrie and I have become big believers in the idea that we should focus on what we have rather than what we don't have. We have weathered some serious storms in our relationship, but on a lot of fronts, we feel closer than ever now. And we really try to live in a way that allows for the word *gratitude*. In fact, Lorrie has since made a career as an outdoors fitness expert, helping other women stay in shape physically and emotionally. As part of her work, she teaches women about the power of accepting life as it presents itself, and enjoying that life.

Lorrie and I have vowed to appreciate each other, appreciate our two daughters, appreciate every day. We don't always maintain that positive attitude. We still have our arguments. But that's our goal.

And so, yes, I choked up seeing our two teenage daughters arm in arm, skipping down that street at Lake Tahoe. It reminded me of what I've missed, and that was hard for me. But it also reminded me of how lucky we all are to have one another, and why we have a duty to try to live happily together, from a place of gratitude.

6

FAST, NEAT, AVERAGE

When passengers are awaiting takeoff on a commercial plane, I'm guessing that most of them don't give a lot of thought to how the pilots in the cockpit got their jobs. Passengers seem most concerned about when they have to turn off their cell phones, or whether it's still possible to use the restroom before the cabin door is closed. They wonder about making their connecting flights, or being stuck in the middle seat. They're not thinking about the pilot's training or experience. I understand that.

Some passengers boarding Flight 1549 at LaGuardia said they had noticed my gray hair, which they equated with experience. But none of them asked about my résumé, my flight record, or my educational background. And why would they? As they should, they

trusted that my airline, US Airways, had rigorously selected its pilots based on federally mandated criteria.

And yet, every pilot has a very personal story of how
he or she ended up in control of that type of aircraft
and in that particular airline's cockpit. We all had our
own unique paths and career progressions, and then
found our way to commercial aviation. We don't often
talk about all the steps we took, even among ourselves,
but every time we pilot a flight, we are bringing with
us all of the things we've learned over the thousands of
hours and millions of miles we've flown.

Until the mid-1990s, 80 percent of pilots working
for major airlines were trained in the military, according to the Federal Aviation Administration. Now, just
40 percent of newly hired pilots get their training in
the military. The rest come through civilian training programs, including some two hundred universities that offer aviation training. The World War II and
Korean veterans—my mentors when I started—retired
as commercial pilots more than two decades ago, after
turning sixty years old, then the mandatory retirement
age. There aren't a great many Vietnam-era pilots left
either, even though the retirement age was raised to
sixty-five in 2007.

As for myself, I am grateful that I came into aviation
through the military. I appreciate the discipline taught

to me during my days in the Air Force, and the many hours of intense training I received. In some civilian programs, pilots aren't always taught with the same rigor.

I was tested in so many significant ways during my time in the service that I sometimes look back and wonder: How did I make it through? How did I succeed when some didn't? How was I able to complete every flight, landing my plane safely, when others I knew and respected didn't make it safely to the runway and lost their lives? As I look back, I reflect on the intersections of preparation and circumstance, and that helps me understand.

My military career provided many of the important steps along the way. The initiation to my military life began in the spring of 1969, when I was a senior in high school and went to see my congressman, Ray Roberts, at his office in the ranching town of McKinney. Then fifty-six years old, he was a well-regarded Democratic leader in Texas, who six years earlier had been in President Kennedy's motorcade in Dallas. He was four cars behind the presidential limousine when the shots were fired.

I had come to Representative Roberts because in order to attend one of the service academies, I'd need a congressional appointment. In some congressional districts, patronage determined which young people got appointments at the Naval Academy in Annapolis, Maryland; the Air Force Academy near Colorado Springs, Colorado; the Military Academy at West Point, New York; the Merchant Marine Academy in Kings Point, New York; or the Coast Guard Academy in New London, Connecticut.

But Representative Roberts believed his appointments should be merit-based. And so he had ambitious young men like me come to his office to be interviewed by a panel of retired generals and admirals who lived in his district.

After we'd gone to the post office to take a civil service exam and scored well enough on it, we were brought before this ad hoc board of military heavyweights at the congressman's office. The board he put together had two tasks. First, to determine whether an applicant had what it took to make it at a service academy. Second, to decide which academy an applicant was best suited for. Since my dad didn't know anyone in high places, I was grateful to have a chance to land an appointment on merit. I had a shot.

I was nervous going to my interview, uncomfortably dressed in my sport coat and tie, but I was excited, too. I'd devoured books about the military and about aviation since I learned how to read. I'd paid attention.

So I was prepared when I finally sat down in front of that panel of four senior officers for their very formal twenty-minute proceeding.

The retired Army general seemed to enjoy lobbing questions. "Mr. Sullenberger," he said, "can you tell me which branch of the service has the most aircraft?"

I assumed most applicants would give the obvious answer: the U.S. Air Force. But I knew this was a trick question. And I had done my homework. I'd studied the specifics of each branch of service and the aircraft they used. "Well, sir," I said, "if you're including helicopters in your count, the U.S. Army would have the most aircraft."

The retired general smiled. I was passing the audition. As we continued talking, he seemed eager to have me go to West Point. But I was pretty straightforward that day. I wanted to fly jets in the Navy or the Air Force, and I didn't want to go to West Point.

As things turned out, Representative Roberts offered the Air Force Academy appointment to another applicant. He gave me the Naval Academy appointment. Then, as fate would have it, the boy with the Air Force slot declined to take it. And so I moved up from the alternate spot.

I was eighteen years old and bound for Colorado. I would be receiving the full ride of a first-class education.

In return, I agreed to pay my country back by serving five years as an active-duty Air Force officer.

I arrived at the Air Force Academy on June 23, 1969, and as a kid from rural Texas, it was an eye-opening moment to meet the other cadets, who hailed from all over the country. Yes, a few of the 1,406 young men in my entering class were wealthy boys from elite families who got there through their fathers' connections. More were sons of military officers, some from families with long military traditions. But once all of us had made our way through long lines to get our heads shaved, it felt as if those distinctions no longer mattered. It would be the same grueling road for all of us. Only 844 of the 1,406 who arrived that day would end up graduating.

We were welcomed to the academy on a gorgeous Colorado morning when there wasn't a cloud in the sky. From that day on, I was amazed by the West. You could see for a hundred miles in any direction, and the mountains were right there—a pretty stunning sight for a boy from the flatlands of North Texas.

The academy grounds were architecturally dramatic, and at the time, the buildings were fairly new. The Air Force Academy had been completed just twelve years earlier; the first class graduated in 1959. If I made it through all four years, I'd be in the class of 1973, which would be just the fifteenth graduating class. The first female cadets wouldn't arrive until 1976, three years after I left.

I was pretty nervous that day. I didn't know what to expect. Unlike some new cadets, I wasn't aware of how intense the hazing was going to be. Once we had put away our street clothes and gotten into our drab, olive-green fatigue uniforms, the upperclassmen showed up and started yelling at us.

"Stand up straight! Suck your gut in!"

"Push your chest out! Get those shoulders back and down!"

"Get those elbows in! Get your chin in, mister!"

"Keep those eyeballs caged straight ahead!"

Did it shake me up? Of course it did. At age eighteen, I lacked the life experiences to put it in perspective. I was a kid straight from my comfortable upbringing, and all of a sudden I was thrust into a situation where I didn't know which way was up. It was disorienting.

It is natural to question the utility of such theatrics. Do I think it was necessary? I'm still not sure. But now as an adult, I do understand some of the rationale for that first-year hazing. It was designed to tear us away from the easy, the comfortable, and the familiar. It was intended to refocus our perspective and reset our priorities. For all of us, it would no longer be about "me" but about "us." That first year began to make real what, until then, had been theoretical constructs—like duty, honor, and "service before self." These words could no longer be thought of as abstractions. Instead they now had real meaning in real life, as in "in-your-face" reality. It's amazing how clearly and how quickly one learns about diligence, responsibility, and accountability when the only allowable, acceptable responses to any query by a superior are "Yes, sir," "No, sir," "No excuse, sir," or "Sir, I do not know."

The rule was that the upperclassmen weren't supposed to get physical with us. But there was some jostling along with the yelling and the intimidation.

Those who argue in favor of hazing say that it builds a sense of loyalty among comrades, and there's some truth to that. As that freshman year wore on, I felt very close to many of my fellow "doolies" (it's a derivation of the Greek word *doulos*, which means "slave"). You have volunteered to fight for your country, and you feel that sense of patriotism. I have heard and read the experiences of those who saw combat, and they say when you get to the battlefield, you're really fighting for your comrades, not some politician or political ideal. You'd rather die than let your comrades down.

My doolie year left me bonded for life with some of my fellow freshmen. It was an intense experience;

it wasn't like just going off to college. We were being tested, abused, physically challenged. And we had to watch a number of those in our ranks fall away. Some wouldn't make it through the mental and physical challenges of basic training. Some would fail academically, or feel too intimidated by the hazing. Others would transfer to regular universities after deciding, "This is not for me. I want a good education, but not at this cost." Those of us who endured and remained became a brotherhood.

That first summer we were sequestered at basic training, and it was the most grueling physical experience of my life. We'd have to go out on formation runs, our rifles held high above our heads, our boots slapping the ground at the same instant, and it was a real sign of weakness or failure to drop out of formation. The upperclassmen would yell: "Keep that rifle high! Don't be a pussy. You're letting your classmates down!"

The guys who had the most trouble were those who couldn't run well enough. They'd get used up and have to stop. And once a cadet dropped out of formation, the upperclassmen would circle him and yell at him. It was very intense. Some men would throw up from the exertion. On rare occasions, someone would cry. Some of my classmates had fathers who were military officers; they feared they'd be disowned if they had to drop out of the academy. I felt for them. I would later wonder where they would end up, at a civilian university perhaps—someplace where you could get a good education without going through all of this.

I had grown up at sea level, and here we were, at an elevation of almost seven thousand feet. It was hard for all of us until we acclimated to the altitude. I was usually somewhere in the middle of the pack, but I held my own. I was determined to make it through the summer, and through the four years to follow.

Though I was homesick and exhausted, I did enjoy some aspects of that summer. They would break us into teams and give us physical problem-solving tests to evaluate us. We were handed a bunch of ropes and boards and, as a team, had to come up with a way to get from one side of a large enclosed cubicle to the other without touching the ground or the water below, and in a limited amount of time. The upperclassmen and officers stood there with clipboards and stopwatches, observing who had the leadership skills to get his team safely across. When it was my turn to be the leader of this exercise, I did pretty well, and that gave me confidence.

I know that summer of training helped me later. It made me realize that if I dug deep enough, I could find strength I didn't know I had. If I hadn't been forced to push myself that summer, I would never have known the full extent of what inner resources I had to draw upon. It wasn't as if I was lazy as a boy. I wasn't. But until that summer, I had never pushed myself to the limit. Those of us who made it through realized that we had achieved more than we thought we could.

When summer was over, the physical demands let up, but the academic demands set in. It was an extensive and difficult core curriculum. No matter your major, you had to take a large number of courses in basic sciences—electrical engineering, thermodynamics, mechanical engineering, chemistry. We also took courses in philosophy, law, and English literature. In retrospect, I am grateful for the education, but at the time, the course load felt staggering.

Luckily, for those of us who so badly wanted to fly, there were just enough perks to keep us motivated.

My first ride in a military jet was during freshman year, in a Lockheed T-33, which dated back to the late 1940s. The plane had a bubble canopy and went about five hundred miles an hour. It was typical of jets from that era; the aerodynamic technology had outpaced the propulsion technology. It was well into the 1950s before jet engines were designed to produce enough thrust to fully take advantage of the strides in aerodynamics.

So this old T-33 was underpowered. Still, it was an incredible thrill to be in it.

Each new cadet was taken for a forty-five-minute ride, and the purpose was to give us an incentive to work hard so we wouldn't drop out of the academy.

This was the first time I'd ever worn a parachute, helmet, and oxygen mask, the first time I had ever been seated on an ejection seat. The officer piloting the plane did a roll, then headed ten miles west of Colorado Springs and flew over Pikes Peak upside down. My stomach was rock solid through all of it. I was so engaged in the moment. I was just eating it all up. I knew that, no matter what, this was what I wanted to do with my life.

When the forty-five minutes were up, of course, it was back to reality. The hazing awaited us on the ground.

We ate breakfast, lunch, and dinner in Mitchell Hall, sitting at rectangular tables of ten. Each table had a mix of freshmen, sophomores, juniors, and seniors. We freshmen had to sit rigidly at attention, our backs straight, our eyes only on our plates. We had to lift our forks to our mouths in a robotic fashion, and we were not allowed to look beyond the food in front of us. We weren't allowed to talk to one another. Only when an upperclassman addressed us, asking us a question, could we speak. They would spend mealtime quizzing us, and we had to shout out our answers.

We each had been given a book called *Checkpoints*, a pocket-size bound volume. We had to memorize all of this legendary lore, and especially the Code of Conduct. When upperclassmen asked us questions, there'd be hell to pay if we didn't know the exact answers.

The Code of Conduct, established by President Eisenhower in 1955, was considered vital because during the Korean War, American POWs had been forced through torture to collaborate. The term in those days was that they'd been "brainwashed." And so the military came up with specific rules of conduct, and we were expected to memorize them all. As future officers, for instance, we had to vow: "I will never surrender the members of my command while they still have means to resist." We could surrender only in the face of "certain death." We had to repeat key lines from the code: "If I am captured I will continue to resist by all means available. I will make every effort to escape and aid others to escape. I will accept neither parole nor special favors from the enemy."

Mealtime became increasingly stressful because the upperclassmen were relentless in their demands. We had to memorize the details about a great number of airplanes. We were expected to know foreign policy,

American and world history, and sports scores from the day before. We had to be able to rattle off the full names of all the upperclassmen at the table, including their middle initials, and their hometowns. Now, forty years later, many of those names and middle initials remain seared into my head. I remember the hometowns, too.

The degree of harassment awaiting you at mealtime depended on your daily table assignment. Walking into the dining hall, if you saw you were seated with a kindhearted senior, you were relieved. But if one of the seniors sitting at your table was a notorious hardass, your heart would sink. You knew dinner would be excruciating.

In that case, you hoped for one of two things: Either another freshman cadet at your table would be so pathetically hopeless at memorization that the upperclassmen would focus on him, which meant they'd leave you alone and you could eat. Or else you hoped that one of your freshman tablemates was a genius or had a photographic memory—someone who got everything right. When upperclassmen came upon a know-it-all, they'd focus all their energies on stumping him, finding the one question he couldn't answer, and then giving him hell for his wrong response. When that happened, the rest of us were ignored and got to eat.

. . .

There was one upperclassman, a year older than I was, who wasn't vindictive about his hazing. But he knew exactly how to make his point.

One day, we were getting ready to march to the noon meal. It was a warm morning and we were in short sleeves. I was standing at attention, and he came up to me, asking if I thought I'd done a good job polishing my black uniform shoes.

"Yes, sir," I said.

"How confident are you?" he asked.

"Sir, I am very confident," I answered. (I wasn't allowed to say: "*I'm* very confident." I had to say "I am." Freshmen were prohibited from using contractions.)

This upperclassman decided to make this into a challenge. "Are you willing to match shines?" he asked. My shoes versus his.

"Yes, sir."

He defined our rules of engagement: "If you are confident that you have a better shine than I do, and it turns out you are right, then I will make your bed tomorrow. If my shoes have a better shine, then you'll make my bed as well as your own."

All of us, including the upperclassmen, had to make our own beds using hospital corners. We had to pull our sheets and blankets tight enough so they wouldn't show any wrinkles. The test was to drop a quarter on the bed. If the quarter didn't bounce, we'd have to pull off all the bedding and start again. It was no fun. So if this upperclassman made my bed the next day, it would be wonderful.

He gave me permission to stop looking straight ahead, and to look down at my shoes and then at his. Our shoes seemed equally shiny. But I chose to be bold. "Sir, I win," I told him.

"Well, it's pretty close," he responded, "but we're not finished yet. Let's compare the soles of our shoes."

He stood on one foot, allowing me to see his instep, the arched middle section between the heel and the ball of the shoe. The leather on each of his insteps had been polished to a sheen. My insteps, of course, were not. He was like a good trial lawyer who never asks a question without knowing the answer. He had set me up.

"Sir, you win," I said. He saw my lips turn into the hint of a grin, and even though doolies weren't allowed to smile while in formation, he cut me some slack by not calling me on it.

There were plenty of other times I had to stifle my smile.

While marching in basic training, we were required to take turns counting off in cadence: "Left, left . . . left, right, left . . ."

Early on in my life, I noticed that accomplished people on TV, especially newscasters such as NBC veterans Chet Huntley and David Brinkley, enunciated perfectly and seemed to have no real accents. I tried to sound more like them, and less like some of the people in my town, who had thick Texas accents. So when it was my job to count off in cadence, I don't think the other cadets could hear Texas in my voice.

But there was one fellow doolie, Dave, who came from West Texas—and you knew it every time he opened his mouth. Whenever he led us, he would count off in cadence: "Lay-uff, lay-uff . . . lay-uff, raah-yut, lay-uff . . ."

I'd chuckle inside, but my face remained expressionless as we marched around. "Lay-uff, lay-uff . . . lay-uff, raah-yut, lay-uff . . ."

We were truly a melting pot at the academy, and sometimes it felt like the clichéd casting in a World War II movie. We had the guy with a Polish name from Chicago, the Texan, the Jewish kid from one of the boroughs in New York, a guy from Portland, Oregon.

It's funny, the things you remember.

When my daughter Kate entered high school in the fall of 2007, Lorrie and I went to back-to-school night, and her math teacher looked familiar to me. As he spoke, it hit me: He was two years ahead of me at the Air Force Academy. He had been one of the upperclassmen asking me mealtime questions my freshman year. So after his presentation, I walked up to him and said: "Fast, neat, average, friendly, good, good." He looked at my face and he had a flash of recognition, too. He knew exactly what I was saying.

At the conclusion of every meal, freshmen at the end of each table had the additional duty of filling out Air Force Academy Form 0-96—the critique of the meal. It was another useless ritual. By tradition, we always filled out the form the same way. How was the service? "Fast." What was the appearance of the waiter? "Neat." How was the portion size? "Average." What was the attitude of dining-room personnel? "Friendly." How was the beverage? "Good." And the meal? "Good."

Kate's math teacher and I shook hands and smiled, two older men recalling the rhythmic, long-ago language of our youth.

In May of 1970, near the end of the freshman academic year, the hazing stopped, and we had what was called "The Recognition Ceremony," formally acknowledging our new status as upperclassmen. That was the day we no longer had to address the older cadets as "sir." We could eat in relative peace. Eventually, when it was my turn to quiz freshmen during meals, I asked questions about flying, as opposed to barking out demands for mindless memorization. I was more comfortable making it educational for the younger guys.

Despite all the regimentation, there was also a sense that your superiors and professors tacitly condoned unauthorized schemes that showed spirit or initiative. Every year, tradition dictated that the freshman class had to assert itself in some way, to prove itself worthy by coming up with antics that equaled or surpassed the stunts tried by previous freshman classes.

Our class seized on the idea of redecorating the outside of the planetarium, where cadets gathered to study astronomy. The large domed building was white like an igloo, but one day well after taps, my classmates sneaked out in the dark of night and covered the building in black plastic, sticking a number eight on the center of the dome. When the entire academy gathered for the march to breakfast, it looked like a huge eight ball. I wasn't involved in the stunt, but I felt a nice charge that day. That and a few other stunts definitely helped our morale.

In the summer before sophomore year, we all endured survival training. We were each sent into the woods for four days without food and water. This was called SERE training, which stood for "survival, evasion, resistance, and escape." It was designed to teach us survival skills, how to avoid being taken as prisoners of war, and how to behave if captured.

The upperclassmen dressed up like communist soldiers and came looking for us. The drama was a bit over-the-top, but it all felt like serious business. I struggled during those days, overwhelmed at times by lack of sleep and food. I was luckier than some of my classmates, because I managed to sneak into the upperclassmen's encampment unnoticed and grab a loaf of bread and some jelly. Others went all those days without eating.

By sophomore year, I realized how much all of these experiences had helped me mature. I had been very homesick in my first six months at the academy. But when I returned home for visits, my homesickness ended. Here I was, not yet out of my teens, but I had met people from all over the world. I had done hard things I didn't know I could do. It was as if I had become a man, and my hometown seemed so much smaller to me than I had remembered.

We hadn't been allowed to fly an aircraft at the academy until the end of our doolie year, so when I got back to Mr. Cook's grass strip on breaks, I was pretty rusty. I didn't have enough flying time to really have the total mind-muscle connection that one has riding a bike. I had to get my bearings again.

Starting in my sophomore year at the academy, I got an amazing amount of flying instruction and experience. I'd get a ride down to the airfield every chance I could.

I also signed up to learn how to fly gliders. I loved flying the gliders because gliding is the purest form of flight. It's almost birdlike. There's no engine, it's much quieter, and you're operating at a slower speed, maybe sixty miles an hour. You feel every gust of wind, and so you're aware of how light your airplane is, and how you are at the mercy of the elements.

Gliding in Colorado, I learned that the way to stay aloft longer is to carefully use the environment to your advantage. The sun heats the surface of the earth unevenly, especially in summer, and so some parts become warmer than others. The air above the warmer parts is heated and becomes less dense, so you have rising air over these areas of the earth. When you fly through a column of rising air, you can feel it lifting the airplane. If you enter a very tight turn to remain in that air, it's like riding an elevator as long and as high as it will take you. It's called "thermal lift," and going from one thermal to another, you can end up soaring for hours.

In the wintertime, you have "mountain wave lift." The winds in the air are stronger in winter, and if the wind is crossing a mountain or ridgeline, it's like water flowing over a rock. If you stay in the rising air down-wind of the mountain, you can remain aloft for long periods.

While I was at the academy, in addition to all the hours spent in gliders, I got my flight instructor certificate. I began to teach other cadets, including a dozen friends, how to fly both airplanes and gliders.

Because I had so much experience, when I graduated from the academy in 1973, I was named "Outstanding Cadet in Airmanship." It was an honor that came because I'd been tenacious in honing my skills through all those hours in the skies.

The Air Force Academy gave me an education on many fronts—about human nature, about what it means to be a well-rounded person, and about working harder than I'd thought possible. On campus, the education we received was called "The Whole Man Concept," because our superiors weren't just teaching us about the military. They wanted us to have great strength of character, to be informed about all sorts of matters we might easily dismiss, and to find ways to make vital contributions to the world beyond the academy. We cadets often dismissed it as "The Manhole Concept," but in our hearts we knew we were held to high standards and difficult tests that would serve us well.

It seemed almost as if the goal was to prepare each cadet to be chief of staff for the Air Force. Only one of my classmates, Norton Schwartz, actually made it; he was appointed to the highest-ranking Air Force job in August 2008. But many of the rest of us did OK, too, in our own way, graduating into the world beyond the academy with a full set of skills and a high sense of duty.

Fast, neat, average, friendly, good, good.

7

LONG-TERM OPTIMIST, SHORT-TERM REALIST

Like my father, I was a military officer who never saw combat. When each of us joined the service, we knew that we might find our lives threatened in war. We soberly accepted that commitment to duty, but neither of us had any visions of martial glory. My father felt honored to serve his country as a naval officer. I saw my years of peacetime Air Force service as a high calling, because every day of training and practice better prepared me to defend my country if called upon.

After spending years readying for tasks they never had to carry out for real, many military men are left to wonder how they would have fared in combat. I understand that, yet I don't feel incomplete because I never saw wartime service. The fighters that I flew were designed to destroy those who would do us harm. I'm glad that I never had to inflict grievous damage on someone else, or to have it inflicted on me.

But I'll never fully know how I would have performed under the pressures of battle. Yes, I faced certain risks on almost every flight I flew as a fighter pilot; it's a dangerous job, even during training missions. Still, over the years, like many who serve in times of peace, I have asked myself questions: If ever faced with the ultimate challenge, a life-or-death moment in battle, would I have been able to measure up? Would I have been strong enough, brave enough, and smart enough to endure the demands of such a test? Would I be able to preserve the safety of those under my command?

My sense is, I would have performed as I was trained. I don't think I'd have panicked or made a grave mistake. But I have accepted the fact that I will never know for sure.

I expected that my commercial airline career would follow a similar pattern. I would take off and land again and again without incident. Yes, airline pilots are trained for emergencies—we practice in flight simulators—and we know the risks, low as they are. The good news is that commercial aviation has made such great strides and is so reliable that it is now possible for an airline pilot to go his entire career without ever experiencing a failure of even a single engine. But one of the challenges of the airline piloting profession is to avoid complacency, to always be prepared for whatever may come while never knowing when or even if you'll face an ultimate challenge.

Because a commercial career can feel routine, I truly didn't think I'd face a situation as dire as Flight 1549. On reflection, however, I realize this: Though I never saw battle, I spent years training hard, paying close attention, demanding a great deal of myself, and maintaining a constant readiness. I survived my own close calls and carefully observed the fatal mistakes made by other pilots. That preparation did not go to waste. At age fifty-seven, I was able to call upon these earlier lessons, and in doing so, answer the questions I'd had about myself.

I graduated from the Air Force Academy on June 6, 1973, and within a few weeks, I enrolled in the summer term at Purdue University in West Lafayette, Indiana, getting my master-of-science degree in industrial psychology (human factors). It's a discipline focused on designing machines that take into account human abilities as well as human limitations. How do humans act and react? What can humans do and what can't they do? How should machines be designed so people can use them more effectively?

It was a cooperative master's program designed to fast-track academy graduates, allowing us to get a graduate degree from a civilian school very quickly, without delaying entry into flight school, which was the next step for many Air Force officers. I had taken graduate-level courses my senior year at the academy, so once the credits were transferred to Purdue, it took me just six more months to get my master's.

At Purdue, I studied how machines and systems should be designed. How do engineers create cockpit configurations and instrument-panel layouts, taking into account where pilots might place their hands, or where eyes might focus, or what items might be a distraction? I believed learning these things could have applications for me down the road, and I was right. It was helpful to get an academic and scientific perspective on the underlying reasons for procedural requirements in flight. When you're learning how to be a pilot, you're often taught the correct procedures to follow, but not always why those procedures are important. In later years, as I focused on airline safety issues, I realized how much my formal education allowed me to view the world in ways that helped me set priorities, so I understood the why as well as the how.

After my six months in Indiana, the Air Force sent me to Columbus, Mississippi, for a year of what is called UPT—Undergraduate Pilot Training. It was a mix of classroom instruction about flying, flight simulator training, and a total of two hundred hours in the air. At first I got to fly the Cessna T-37, which is a basic twin-engine, two-seat trainer aircraft used by the Air Force. It was twenty-nine feet long with a maximum speed of 425 miles per hour. Eventually, I graduated to the Northrop T-38 Talon, which was the world's first supersonic jet trainer. It could reach a maximum speed of over 800 miles an hour, which is more than Mach 1.0.

I'd come a long way from the days of slowly circling Mr. Cook's field in his Aeronca 7DC propeller plane, barely topping a hundred miles an hour. Now I was being taught skills that would allow me to fly at high speeds in formation, my wings just feet away from the jets on either side of me. And I was sitting on an ejection seat, ready to bail out if my jet became unflyable.

I was twenty-three years old then and my two instructors in the T-37 and then the T-38, both first lieutenants, were a few years older. They were from Massachusetts and Colorado, and they had something wonderful in common: They weren't just teaching me because they were required to do so. "I want you to succeed," they each told me, and they offered every bit of guidance they could give me.

After Mississippi, the Air Force sent me to Holloman Air Force Base near Alamogordo, New Mexico, a base with a storied history. During World War II, it had served as the training ground for men flying Boeing B-17 Flying Fortresses and the Consolidated B-24 Liberator, which was the most common heavy bomber used by Allied forces.

The B-24 was designed to have a long range, and more than eighteen thousand of them were manufactured quickly during the war. But flight crews found that the plane was too easily damaged in battle, and given a design that placed fuel tanks in the upper fuselage, it was too likely to catch on fire. The B-24s delivered their payloads—each plane could hold eight thousand pounds of bombs—but a lot of lives were sacrificed in order to do so. Many of those lost men passed through Holloman before me.

Holloman was known for other historical achievements, too. On August 16, 1960, Captain Joseph Kittinger Jr. took an open balloon gondola to 102,800 feet to test the feasibility of high-altitude bailouts. He stepped out of the balloon over Holloman and fell for four minutes and thirty-six seconds, at a velocity of 614 miles an hour, the longest free fall a human being had ever endured. His right glove malfunctioned, and his hand swelled to twice its normal size, but he survived and was awarded the Distinguished Flying Cross.

Like Holloman, every base where I was stationed had a history that inspired me. It was almost as if you could feel the presence of heroes in the winds over the runways.

I was at Holloman for "FLIT," which stood for "fighter lead-in training." We worked on basic air combat maneuvers, tactics, and flying formation in the T-38. I knew I wasn't a true fighter pilot yet, but training at Holloman, I knew I was going to be. I had a lot to learn, but I had the confidence that I could do it.

You couldn't avoid the feeling that you were in elite company. There had been thirty-five men in my pilot training class in Mississippi. Many of them wanted to fly fighters. Just two of us were chosen to do it. So I took seriously that my superiors had faith in me, and I worked hard at Holloman to live up to their expectations.

Next stop was a ten-month stint at Luke Air Force Base near Glendale, Arizona, where I checked out on the F-4 Phantom II. The supersonic jet, which can fire radar-guided missiles beyond visual range, flies at a maximum speed of over 1,400 miles an hour, or Mach 2.0. Unlike many fighters, the F-4 was a two-seat airplane. The pilot sat in the front seat and a specially trained navigator called a Weapons Systems Officer (WSO) sat in the back seat.

We went through the F-4 system by system—electrical, hydraulics, fuel, engines, flight controls, weapons, everything. We looked at each system individually and how they worked together as a whole.

My fellow pilots and WSOs and I learned not just how to fly the F-4—that was the easy part—but how to use it as a weapon. We dropped practice bombs. We engaged in air-to-air combat training. We practiced flying in tactical formation. We also learned to work closely with our WSOs as an effective team.

Day after day, we learned the intricacies of the machine, and learned about our abilities or inabilities to master it. And equally important, we learned a great deal about one another.

This kind of flying was very demanding and exciting at the same time. So much of what we had to do in the cockpit was manual. We didn't have the automation that exists today to help us figure out things. Unlike those who pilot current fighters, with complex computerized systems, we had to do most everything visually. Today, computerization enables flight crews to release bombs that hit targets with pinpoint accuracy. In the older fighters that I flew, you had to look out the window and make estimations in your head. Before you flew, you'd go over the tabulations of numbers, determining when you'd have to release a bomb given a certain dive angle, speed, and altitude over the target. If you were slightly shallow or steep in the dive angle, the bomb would go short or long. In a similar fashion, the speed at release and the altitude at release also affected whether the bomb would go short or long. You also had to allow for crosswinds when you flew over the target. Modern airplanes provide pilots with far more guidance about how to do all these things precisely.

In 1976 and early 1977, I spent fourteen more months flying the F-4 while stationed at Royal Air Force Lakenheath, seventy miles northeast of London. It was my first assignment as an operational fighter pilot.

Jim Leslie, now a captain with Southwest Airlines, was a contemporary of mine in the Air Force. We arrived at Lakenheath within a few days of each other back in 1976, and we looked a lot alike. We were both skinny, six-foot-two blond-haired guys with mustaches. When we showed up together, people would get us mixed up. Some didn't even realize we were two separate guys until they saw us in the same room.

A lot of the older pilots knew one of us was named Sully, but they weren't sure at first which one of us it was. "Hey Sully!" they'd say, and after a while, Jim got so used to being addressed that way that he'd turn around, too. When I landed Flight 1549 in the Hudson, I'm guessing there were some old fliers from the Lakenheath days who pictured Jim as the "Sully" at the controls.

By his own admission, Jim was a bit of a hot dog in the skies. I had the predictable call sign of "Sully." His call sign was "Hollywood," and he wore fancy sunglasses and unauthorized boots that were part cloth, part leather. He was a bit flamboyant, but he was also smart and observant. He'd put things in perspective. As he liked to say it: "It's impossible to know every last bit of technical stuff about how to fly fighter planes, but we ought to know as much as we can because we need to be the go-to guys."

After Lakenheath, I did a three-year stint at Nellis Air Force Base in Nevada, where I rose to the rank of captain. Jim was stationed there, too.

He and I became close, though we took different approaches as aviators. He took pride in being a bit of a loose cannon. I considered myself more disciplined. When we were dogfighting, there were rules for how far away you had to be from another jet when you passed it head-on. If the instructions were that we get no closer than a thousand feet, Jim would try five hundred feet. "I know I can do it," he'd say, and he was right. "Sully, you can do it, too." I knew I could, but I knew that if I did, I'd be shaving the margins we needed in order to avoid the unexpected, when a slight misperception or misjudgment could put two airplanes too close.

I respected Jim. He knew he wasn't really putting anyone in danger, because he knew his own skills. But this was training, not combat. I was more judicious in my use of aggressiveness. There would be times in my career, including my years as a commercial airline pilot, when it would be useful and appropriate to use a bit of aggression.

The bonds among pilots were paramount. At each base where I was stationed, we were reminded again and again how vital it was to know about the dangers of complacency, to have as much knowledge as possible about the particular plane you were flying, to be aware of every aspect of what you were doing. Being a fighter pilot involved risk—we all knew that—and some accidents happened owing to circumstances beyond a pilot's control. But with diligence, preparation, judgment, and skill, you could minimize your risks. And we needed one another to do that.

Fighter pilots are a close-knit community in part because it's necessary for everyone's survival. We had to learn to take criticism and also how to give criticism when needed. If a guy makes a mistake one day, you can't ignore it and let it pass. You don't want him making the same mistake the next time he flies with you. You've got to tell him. Your life, and the lives of others, depend on it.

I'm guessing I met five hundred pilots and WSOs in the course of my military career. We lost twelve of them in training accidents. I grieved for my lost comrades, but I tried to learn all I could about each one of their accidents. I knew that the safety of those of us still flying would depend on our understanding of circumstances when some didn't make it, and our internalizing the vital lessons each of them could leave as a kind of legacy to us, the living.

America labeled Charles Lindbergh as "Lucky Lindy," but he knew better. I've read *We*, his 1927 book about his famous transatlantic trip. In it, he made clear that his success was due almost entirely to preparation, not luck, or as I prefer to call it, circumstance. "Prepared Lindy" wouldn't have had the same magic as a nickname, but his views of pilot preparation have long resonated with me.

Whenever a fellow airman lost his life during my military career, I tried to think of how I might have reacted, and what steps I might have taken. Could I have survived?

At Nellis, each pilot and his WSO were assigned a particular airplane. We had our names stenciled on the canopy rails.

At one point, I was on temporary duty (TDY) at Eglin Air Force Base in Ft. Walton Beach, Florida. I was there to have a rare opportunity to fire an air-to-air missile at a remotely controlled target drone over the Gulf of Mexico.

One morning while I was in Florida, another crew was scheduled to fly my plane, an F-4, back at Nellis. The F-4 had a nosewheel steering system that was controlled electrically and powered hydraulically. There was an electrical connector that had wires to connect the cockpit control with the nosewheel. Once in a while, moisture would get into the connector. If there was contamination there, it would short out the connector pins. So the nosewheel could end up turning without a command from the pilot. We'd have to write it up in the aircraft maintenance log and the technicians would check it and repair it as necessary. Sometimes, the connector simply needed to be dried out so it would work properly.

This pilot was set to take my plane for a training flight that morning. Taxiing to the runway, he noticed that the nosewheel steering was not working properly. He taxied back to the ramp, shut the airplane down, and reported the discrepancy in the maintenance log. The maintenance crew took corrective action and signed it off.

Later that day, that same F-4 was scheduled for a flight, including a formation takeoff, in which pilots in two jets were going to power up, release their brakes, and then take off in formation, matching each other exactly in acceleration.

One of the pilots in the formation was at the controls of the F-4 that had been assigned to me, the one which had aborted earlier in the day. After he started his takeoff, the nosewheel turned sharply to the left without him commanding it to do so. It took him into a ditch beside the runway, collapsing the landing gear and rupturing one of the external fuel tanks.

He and his WSO were sitting in the damaged airplane, deciding how to extricate themselves, when the leaking fuel caught on fire, and they were engulfed in a ball of flames.

I wonder if I had been the next pilot to fly that plane, would I have read the maintenance record, seen how the nosewheel issue had been addressed, and known to be especially watchful of any evidence that it would fail again?

That pilot and WSO were a good crew. But at their funerals, I was reminded of how a crew must be diligent on every front on every flight.

This was graphically illustrated in my own close calls, too.

One time at Nellis, I was in an F-4, on a high-speed, low-level flight. The goal was to fly as low as possible, which is what I'd need to do if I ever had to fly below enemy radar. I was flying just a hundred feet off the ground at 480 knots, and there were hills I had to go over. The techniques I was practicing required me to maneuver the jet so it would barely clear a hill without getting too high above it. Flying too high would make me obvious to enemy radar.

Doing this properly took a lot of practice. Each time I had to raise the nose to fly over the hill, and then push the nose back down after we'd cleared the hill. It was a bit like riding a roller coaster. If it was a steeper ridgeline, I'd come up to it, pull up steeply to clear it, and as I crested the top, I'd roll inverted, upside down, and then pull down the back side of the hill and finally roll back to the upright position.

At one point, I came to a ridgeline, thought it was tall enough that I would be able to pull up to the crest, roll inverted, and pull back down the back side. At the top of the ridge, I realized I didn't have enough altitude ahead of me to complete the maneuver. It was a potentially fatal misjudgment on my part. I had to quickly push myself back up into the sky and then roll out.

I had seconds to correct the situation, and I managed to do it. But let me tell you: The incident got my attention. There were pilots who died after making similar misjudgments.

When we got back into the squadron building, I took responsibility for what happened. I turned to the WSO who had been with me and said, "I'm sorry, Gordon. I almost killed us today, but it won't happen again." I then explained to him exactly what had happened and why.

After I had been at Nellis a few years, I was assigned to an Air Force Mishap Investigation Board. We investigated one accident at Nellis in which a pilot in an F-15 had tried an aggressive turning maneuver too close to the ground. He didn't have enough room to complete it. The vast desert practice ranges we used, north and west of Las Vegas, had elevations starting at about three thousand feet above sea level and going much higher. If you're looking at your barometric altimeter, it is set to give a reading of your altitude above sea level. It doesn't give you the height above the surface of the ground. The pilot apparently misjudged how high he was, and how much room he had. It's likely he realized this when it was impossible to correct. As pilots say, "He lost the picture." Errors had crept into his situational model and had gone unnoticed and remained uncorrected until too late.

I had to take statements from the other guys in his squadron. I had to sort through photos of the accident, shots of the shattered plane on the desert floor. The carnage was chronicled in exact detail, including photos of identifiable sections of the pilot's scalp.

As in every Air Force accident, investigators had to turn all the specific circumstances inside out to learn exactly what transpired. It was as if the pilot who died still had a responsibility to help ensure the safety of the rest of us, his fellow aviators.

Pilots are taught that it is vital to always have "situational awareness," or "SA." That means you are able to create and maintain a very accurate real-time mental model of your reality. Investigating this pilot's apparent inaccurate SA reminded me of what was at stake for fighter pilots. It took an absolute commitment to excellence because we were required to do incredible things close to the ground and fast, often changing directions quickly, while always making sure that the way we were pointed was safe to go.

In so many areas of life, you need to be a long-term optimist but a short-term realist. That's especially true given the inherent dangers in military aviation. You can't be a wishful thinker. You have to know what you know and what you don't know, what you can do and what you can't do. You have to know what your airplane can and can't do in every possible situation. You need to know your turn radius at every airspeed. You need to know how much fuel it takes to get back, and what altitude would be necessary if an emergency required you to glide back to the runway.

You also need to understand how judgment can be affected by circumstances. There was an aircrew ejection study conducted years ago which tried to determine why pilots would wait too long before ejecting from planes that were about to crash. These pilots waited extra seconds, and when they finally pulled the handle to eject, it was too late. They either ejected at too low an altitude and hit the ground before their parachutes could open, or they went down with their planes.

What made these men wait? The data indicated that if the plane was in distress because of a pilot's error in judgment, he often put off the decision to eject. He'd spend more precious time trying to fix an unfixable problem or salvage an unsalvageable situation, because he feared retribution if he lost a multimillion-dollar jet. If the problem was a more clear-cut mechanical issue beyond the pilot's control, he was more likely to abandon his aircraft and survive by ejecting at a higher, safer altitude.

My friend Jim Leslie was on a training mission in an F-4 in 1984, dogfighting with other airplanes. His plane ended up in a spin due to a mechanical malfunction, and there was no way to get it to fly again. "Pilots are only human," he later told me. "In stressful situations, your brain tells you what you want to hear and see, which is: 'This ain't happening to me!' And so you mentally deny that your plane is going down. You think you have time to fix the problem or to escape, when really, you have no time. And so you eject too late."

Jim pulled his ejection handle, which first sent his WSO out of the F-4, then sent him a split second later. "I thought I had ejected us in plenty of time," he said, "but I later learned that I did it just three seconds before the plane hit the ground." Had he waited even one second longer, he wouldn't have made it safely out of the aircraft.

"Nobody wants to crash," Jim said. "It's not a good mark on your flight record. The loss of that F-4 cost the Air Force four million dollars that day. But I lived. And some people die because they don't want to be responsible for the cost of the plane."

Jim later had a chance to fly the F-16. Two of his roommates died in F-16 training accidents, and the job fell to Jim to pack up their gear and return it to their families. Later, Jim would again have to eject from an unflyable plane, an F-16. Again, he survived. "Every day I wake up is a bonus," he'd tell me.

Perhaps the most harrowing flight of my military career came in an F-4 out of Nellis. My "GIB" ("guy in back" or "backseater") was Loren Livermore, a former bank clerk from Colorado who decided to abandon his desk job and become an Air Force navigator. He and I were on a gunnery range over the Nevada desert. I was leading a formation of four fighters, flying a box pattern around the target on the desert floor as part of bombing practice.

We were at a very low altitude, and I felt the plane move by itself. Imagine being in your car, driving along, and all of a sudden, without turning the steering wheel, you start veering to the left. It would be a bit shocking.

For us, in the F-4, the unsettling moment came when we felt the plane make a sudden uncommanded flight control movement.

Loren had hooked up a cassette recorder so he could have a record of what we said to each other, and of our radio transmissions. My response to this movement was very clear on the tape.

"Goddamn it!"

"What was that?" Loren shot back.

"I don't know," I told him.

Being just a hundred feet above the ground, traveling 450 knots, in a plane with a mind of its own—that's not a path you want to be on. I immediately pulled the F-4 skyward. I needed a rapid climb to get away from the unforgiving ground. I had to buy myself time and give myself room. At a higher altitude, Loren and I might be able to make sense of the malfunction and deal with it more effectively. More important, if the situation worsened, we would have the time and altitude to be able to recover, or successfully eject and survive.

I radioed, "Tasty one one, knock it off." That was my order to the other three planes to abandon the practice run and stop the training mission.

Each pilot acknowledged my order.

"Two knock it off."

"Three knock it off."

"Four knock it off."

"Mayday! Mayday! Mayday!" I said. "Tasty one one. Flight control malfunction."

As leader of the formation, I still had to give direction to the other three planes. "Two and four go home," I said. "Three join on me."

I wanted two of the jets to go back to Nellis. They could serve no useful purpose, and I didn't want the increased workload of being responsible for them anymore. As flight lead, I had a responsibility to my flight of four jets as well as myself and my WSO. It was prudent to stop the training when it was no longer reasonably safe and to focus my attention on the higher priority of merely staying alive a little longer.

I chose to have No. 3 escort me, since he was also a flight lead and was more experienced than either No. 2 or No. 4. I wanted No. 3 to see if he could help me make sense of whatever my F-4's malfunction was. Before 2 and 4 left the range, and the frequency, I radioed, "Tasty one one, armament safety check complete."

Each of the other pilots responded.

"Two, armament safety check complete."

"Three, armament safety check complete."

"Four, armament safety check complete."

This ensured that all arming switches were returned to the safe position before planes left the range.

The No. 3 pilot was George Cella. At the time there was a popular TV commercial for Cella Lambrusco wine. The lovable character in the commercial, named Aldo Cella, was a short, pudgy Italian guy with a dark mustache. He wore a white suit and hat, and had women hanging all over him because of his brand of wine. So George's tactical call sign was "Aldo."

Aldo said, "Better do a controllability check."

When I got to a higher altitude, about fifteen thousand feet, I slowed down the jet to make sure it would remain controllable at a slower speed when the time came for me to attempt a landing. Loren, my WSO, turned to the appropriate troubleshooting page, E-11, in our emergency checklist and we verified we could control the plane.

Aldo flew his jet very close to mine. He and his WSO inspected the exterior of my aircraft, looking for any obvious damage, fluid leaks, or other anomalies. "You look OK," Aldo said as he chased me in his F-4.

I contacted Las Vegas Approach Control and advised the civilian controller of my emergency status and of my need to return for landing at Nellis. The controller put certain constraints on how I might return, and how long I could take to line up. He wanted a tighter turn to my final approach.

"Unable," I told him. That's the standard response when a pilot can't do what a controller is asking him.

I told him I needed a five-mile final approach to make sure I could be stabilized for landing. I was glad I had insisted on that, because as I was descending, a gust of wind caused a wing to dip. Aldo and his backseater assumed I was losing control of the F-4. They expected to see Loren and me flying like cannonballs out of our plane in our ejection seats. But I moved the control stick full right, and was able to raise the left wing that had dipped. For the moment, we held on.

After that gust of wind, I was intensely focused on keeping the wings exactly level, and on carefully maintaining both our vertical and horizontal path to the runway. I tried to get exactly in line with the runway's centerline.

Aldo followed me down, ready to let me know the instant I deviated from the proper path or entered an attitude from which I couldn't recover. I felt like I was still in control, but I was wary, prepared for the possibility that my aircraft might betray me and I'd have to abandon it.

We made it over the safety area leading up to the runway threshold, and within a few seconds, we were on the runway itself, our drag chute deployed.

We had made it safely to the ground.

I braked to a stop, then slowly taxied back to where the other fighters were parked. Loren and I stepped off the ladder, and stood there for a moment. We were both holding our helmets and oxygen masks in our left hands, but our right hands were free. Loren reached out to shake my hand, and said, from his heart but with a big grin, "I thank you, my mother thanks you, my brother thanks you, my sister thanks you . . ."

Loren and I had worked together as a team, with help from Aldo and his WSO. We had maintained control of the aircraft and solved each problem so we could land safely.

Had I died that day, other pilots would have grieved for me. Fellow pilots would have been assigned the duty of investigating the accident. They would have learned the cause of my crash. I'm glad I saved them from having to look at a photograph of my scalp.

Each man we lost had his own regrettable story, and so many of the particular details remain with me.

At Nellis, there was Brad Logan, my "wingman" (which meant he flew the aircraft beside me, following my lead). There would be four planes in formation, and Brad was in the number two plane. We flew together more than forty times. He was a very good pilot.

I was a captain, and he was a first lieutenant, a few years younger than I was. He was an unpretentious, unassuming, jovial guy who was always smiling. Big, solid, and friendly, he looked like Dan Blocker, the actor who played Hoss Cartwright on *Bonanza*. Naturally, Brad's tactical call sign was "Hoss."

After Nellis, he was flying out of an air base in Spain. One day, on a training mission, his plane was in formation descending through the clouds. I heard there was a miscalculation or miscommunication between air traffic control and the leader of his flight. Maintaining his assigned position in the formation, through no fault of his own, Brad's plane crashed into the side of a mountain obscured by clouds. The other planes in the formation were high enough to fly over the mountain, but Brad and his backseater were killed.

He had a wife and a young child, and as I recall, they received just $10,000 or $20,000 from his government life insurance policy. That's how it was for pilots' families after their accidental deaths; the support they received was very modest. But we signed up knowing this. We were aware that some of us wouldn't make it because not all training exercises could go flawlessly. There was always the chance that surprises such as low clouds and an unexpected mountain could be our undoing.

Those who survived accidents often found ways to acknowledge to the rest of us that they had cheated an unkind fate. They had a bit of an aura about them.

There was a terrific pilot named Mark Postai who was stationed with me in England in 1976. He was a very smart, skinny guy in his mid-twenties, with dark hair and an olive complexion. He had majored in aeronautical engineering at the University of Kansas.

On August 14, 1976, Mark took off from runway 6 at RAF Lakenheath, heading to the northeast, and there was a thick forest off the end of the runway. He had a flight control malfunction that made the airplane unflyable, but he and his backseater were able to eject successfully before the plane crashed into the forest and exploded in a fireball. They survived, uninjured.

When Mark made it back to the base, someone told him: "You know that forest belongs to the Queen of England."

He replied, with a smile, "Please tell the Queen I'm sorry I burned down half of her forest."

Mark lived in the officers' quarters assigned to bachelors, and a week or so after the accident, he invited us into his room for a party. "I want you guys to see something," he told us.

Air Force personnel had searched the woods and found the ejection seat that had saved his life. In appreciation, Mark had put it on display in the corner of the room. "Go ahead, sit in it," he said. We all had drinks in our hands—there was a nurse from the base in the room with us also, I recall—and it just seemed like a very appropriate thing to do, to plant ourselves in that seat and feel the magic. Maybe it offered us reassurance that an ejection seat might save our lives someday also.

Mark told us how it felt to eject, how his heart was pounding. We all knew the science behind ejection seats, of course. A sequence of events must happen to get you out of the jet. Once you pull the ejection handle, the canopy flies off. Then there's a ballistic charge, which is similar to a cannon shell that catapults you out of the airplane. And once you get a certain distance from the aircraft, a rocket motor sustains you and keeps you moving with a slightly more gentle acceleration. After the rocket finishes firing, the parachute deploys itself. The seat falls away, and you parachute down to the ground.

That's if all goes well, as it did for Mark.

The night of his party, he proudly showed us the letter he had received from Martin-Baker Aircraft Company Ltd., which billed itself as "a producer of ejection and crashworthy seats." Evidently, they sent one of these letters to every pilot who had used one of their seats and lived. In the letter, they told Mark: "You were the 4,132nd person to be saved by a Martin-Baker ejector seat." (The British say "ejector" instead of "ejection.")

Like me, Mark's next assignment back in the States was at Nellis, flying the F-4. Because of his skill as a pilot, and his engineering training, he was asked to be in a special "test and evaluation" squadron. The group operated in great secrecy. I figured he was flying stealth fighters.

Mark ended up marrying a young and very attractive woman named Linda. His life was coming together. And then one day, we got word that he had died in an accident. None of us knew what kind of plane he had been flying, but we were told that his death resulted from, of all things, an attempted ejection that had failed.

Only recently, more than two decades later, did I learn through the aviation magazine *Air & Space* what had happened to Mark. The article offered a look at how the United States worked to get inside knowledge about enemy planes during the cold war, especially Soviet MiGs. The story briefly touched on an American pilot who died ejecting from a MiG-23 in 1982. It was Mark. Turned out, the plane had come into American hands somehow. Mark's job was to train U.S. fighter pilots to be able to fight effectively against Soviet aircraft.

The article mentioned a book, *Red Eagles: America's Secret MiGs,* which I tracked down. The book explained that the single engine in the MiG that Mark was flying caught on fire. He began an attempt at an engine-out landing at his desert base but had to eject.

The Soviet fighters had ejection seats with notoriously bad reputations. I assume Mark knew this when he pulled the ejection handle and hoped for the best.

Very few pilots ever have to eject once in their lives. My long-ago friend Mark ejected twice. The second time, of course, there was no congratulatory letter waiting for him from the company that made the ejection seat.

A couple of years after Mark died, I found myself at a social event where Linda, his young widow, happened to be. I told her that I thought her husband was a terrific guy and a gifted pilot, and that I had always enjoyed his company. I told her how sorry I was. And then I was quiet. There wasn't much more I could say.

I guess I felt like something of a survivor by 1980, as my Air Force career was ending. No, I had never been in combat. But unsettling things happened just often enough to get my attention. I knew what was at stake.

There were a dozen different ways on a dozen different days that I could have died during my military years. I survived in part because I was a diligent pilot with good judgment, but also because circumstances were with me. I made it to the other side with a great respect for the sacrifices of those who didn't. In my mind, I can see them—young, eager faces that are with me still.

8

THIS IS THE CAPTAIN SPEAKING

Military units from all over the world came to Nellis to use the endless miles of open Nevada desert to practice maneuvers. I flew against not just the Marines and the Navy but also the Royal Air Force from Great Britain, and units from as close as Canada and as far away as Singapore.

Nellis is famous as the home of "Red Flag," which meant that three or four times a year, we'd engage in weeks-long war games and exercises. We'd be split up into "good guys" and "bad guys" and then we'd take to the skies, devising tactics to fool our adversaries and avoid getting shot down.

Red Flag began in 1975 as a response to deficiencies in the performance of pilots new to combat during the Vietnam War. An analysis by the Air Force, dubbed

"Project Red Baron II," found that pilots who had completed at least ten combat missions were far more likely to survive future missions. By the time they had ten missions under their belts, they had gotten over the initial shock and awe of battle. They had enough experience to process what was going on around them without being too fearful. They had enough skill and confidence to survive.

Red Flag gave each of us "realistically simulated" air-to-air combat missions, while allowing us to analyze the results. The idea was this: Give a pilot his ten missions, and all the accompanying challenges, without killing him.

We were able to have dogfights over thousands of square miles of empty desert. We could drop bombs and go supersonic without bothering anyone. We had mock targets—old, abandoned tanks and trucks—out there. Sometimes we'd drop dummy bombs and sometimes we'd use live ordnance, and we'd have to make sure everyone in formation was far enough away so shrapnel from the bomb explosion wouldn't hit anyone's plane.

Each jet had a special instrument pod that electronically recorded what was going on. There was radar coverage in the desert to monitor attacks, and whether the shots taken were valid. We'd have mass briefings before the exercises and mass debriefings afterward.

On one mission, I was given the opportunity to be the Blue Force mission commander, responsible for planning and leading a mission involving about fifty aircraft. It was a complicated task, planning high-speed, low-level attacks using different kinds of airplanes. We had to figure out when to attempt midair refuelings, how to avoid threats, and how best to use all the available resources to achieve the best outcome. It took leadership and coordination skills, getting everyone on the same page.

Exercises such as Red Flag were thrilling, but other aspects of military life were less appealing to me.

As I approached the end of my service commitment in the late 1970s, I got the sense that the best part of my military career was already behind me. I'd served six years and I just loved flying fighters. But I had learned that if I wanted to have a successful, rising career as an Air Force officer, I'd have to do a lot more than climb into a cockpit and fly. To keep getting promoted, I'd have to choose a career path that took me further away from flying. I'd have to spend much of my time giving briefings or sitting at a desk, signing off on paperwork.

In the peacetime Air Force, appearances mattered. Not just haircuts and shoeshines, but also how you appeared to those above you in the hierarchy. To get promoted, you had to be a good politician. You needed to develop alliances and find a well-connected mentor.

Yes, certain people respected my flying abilities, but I was never particularly good at networking. I didn't put the effort into it. I felt I could get by on my own merits as an aviator.

There were other things that also factored into my decision to leave the Air Force. By the late 1970s, with the Vietnam War over, there was a big drawdown in the military budget. The cuts were exacerbated by rising fuel costs, which meant that to save money, we weren't being permitted to fly as much. It takes years to get good at using a jet fighter as a weapon, so it was crucial to get pilots into the air as often as possible. The budget issues would leave me grounded more than I would have liked.

My career decisions at that time in my life had a lot to do with the simple question: How much will I get to fly?

The idea of applying to be an astronaut certainly had great appeal to me, but by the late 1970s, when I might have tried to qualify, manned missions weren't in the forefront of NASA's plans. The Apollo program, which had sent twelve men to the moon between 1969 and 1972, had been canceled. The space shuttle wasn't yet in operation. Two of my academy classmates would end up flying the space shuttle in the early 1990s, and in many ways I envied them. But I knew I'd have to spend years and years of my life preparing to fly just once or twice in space. That's if I could even have made the cut. I didn't have an engineering degree, and had never been a test pilot as my two classmates had been.

My last day of military service was set for February 13, 1980, three weeks after my twenty-ninth birthday. President Reagan had just taken office and the hostages had been released in Iran. It looked like the nation was headed into more peaceful times, and it felt like the right time for me to return to civilian life.

My final flight was an air-to-air combat training mission, and as you can imagine, it was bittersweet. I flew against our squadron commander, Lieutenant Colonel Ben Nelson, and we both knew the emotions I was feeling at the controls. After the flight, I climbed out of the jet, shook hands with Lieutenant Colonel Nelson and some other well-wishers on the ramp, and then I gave a final salute. It was a simple good-bye.

"Good luck, Sully," Lieutenant Colonel Nelson said.

It was official. I would never again fly a fighter. That's not to say I wasn't a fighter pilot, though. Just as there's no such thing as an ex-Marine, I would always be a fighter pilot.

. . .

I sent an application to almost every airline, but it was not an easy time to get a job as a commercial pilot. The airlines were losing money and starting to feel the effects of federal deregulation fifteen months earlier. There were growing issues between management and labor. In the decade to follow, more than a hundred airlines would go out of business, including nine major carriers.

All of the airlines combined hired just over a thousand pilots in 1980, and I was grateful to be one of them. I came cheap, too. When I started at Pacific Southwest Airlines, as a second officer/flight engineer on the Boeing 727, I was earning less than $200 a week. That was my gross, not my take-home pay.

There were eight of us in my PSA class of new hires, and I rented a room in San Diego with a former Navy pilot named Steve Melton. Steve and I went to class all day, training to be flight engineers. We later had simulator training, after which we would return home and turn our closet into our own little makeshift cockpit. On the inside of the closet door, we taped posters with mock-ups of a flight engineer's panels. We quizzed each other on every light, dial, switch, and gauge, and all the procedures we had to know. We had a lot to learn, and little time to do it.

All eight of us in my class of new hires were so broke that on a lot of afternoons, we'd go to an aviation-themed restaurant-bar close to the airport. The place served one-dollar beers during happy hour, and appetizers were free. That would be our dinner several nights a week.

I entered the airline industry at the tail end of what's been called the Golden Age of Aviation. Before deregulation, flying was relatively more expensive, and for a lot of people, it felt like a special occasion when they went to the airport to fly somewhere. When I arrived in 1980, everything had gotten a little more casual, but you still saw a lot more men and women in dress clothes than you see today. These days, a growing percentage of travelers look like they're on their way back from the gym or the beach or just working in the yard.

Airline service was a lot more civil and accommodating back when I started. On most major airlines, whether you were in first class or coach, you got a meal. Children flying for the first time were given wings and tours of the cockpit. Flight attendants would even ask passengers if they'd like a deck of playing cards. When was the last time you were offered playing cards on an airplane?

From the start, I was very happy to be an airline pilot. True, I had honed skills I no longer needed.

I wasn't going to have to refuel my aircraft in-flight from another aircraft. I wouldn't be dropping any bombs or practicing aerial combat. I wouldn't have to fly at a hundred feet above the ground at 540 knots. But I appreciated being given the opportunity to join such a prestigious profession—one that only a few people get to join, but that many would have liked to.

It's interesting. After you fly for an airline for a while, you realize that it doesn't really matter what your background is. You could have been the ace of your base, or even a former astronaut. You could have been a war hero. Your fellow pilots might respect you for that, but there's no real impact on your career. What matters most is your seniority at that particular airline. How many years have passed since you were hired? The answer to that decides your schedule, your pay, your choice of destinations, your ability to decline flying red-eyes, everything.

Over the course of my career, working harder or being more diligent didn't lead to faster promotions. I spent three and a half years as a flight engineer, followed by four and a half as a first officer. After my eighth year at PSA, I checked out as a captain. My advancement came fairly quickly, but it wasn't because my competence was being recognized. It was because my airline was growing at the time, enough people senior to me were retiring, and enough new airplanes were joining the fleet, necessitating more captains. I was OK with how my promotion was decided.

I also understood the history behind our profession's dependence on a seniority system. It started in the 1930s, as a way to avoid the favoritism, cronyism, and nepotism rampant in the early days. It was about safety as much as fairness. It insulated us from office politics and threats to hinder our careers if we didn't "play the game." A layman might think such a seniority system would lead to mediocrity. Nothing could be further from the truth. Pilots are a pretty proud bunch and they find it rewarding when they have the respect of their peers. The system works.

What the seniority system does not do is afford lateral mobility. We are married to our individual airlines for better, for worse, for richer, for poorer, until death do us part (or until we get our last retirement check).

When you share a cockpit with another pilot, even before you leave the gate, you notice things. You can tell how organized a pilot is, his temperament, his interests. What ways has he found to handle the distressing and the distracting issues of pay cuts and lost pensions, which all of us now face? How does he interact with the flight attendants, especially if his ex-wife used to be one?

After you fly with him for a while, you build on your impressions. Everyone I fly with is competent and capable. That's basic. But is the guy in the next seat someone I can learn something from? Does he have such skill that he makes everything look easy (when we all know it's not)?

Pilots I have known who make it look the most effortless have something that goes beyond being competent and beyond being someone who can be trusted. Such pilots seem able to find a well-reasoned solution to most every problem. They see flying as an intellectual challenge and embrace every hour in the sky as another learning opportunity. I've tried to be that kind of pilot. I've derived great satisfaction from becoming good at something that's difficult to do well.

Before I go to work, I build a mental model of my day's flying. I begin by creating that "situational awareness" so often stressed when I was in the Air Force. I want to know, before I even arrive at the airport, what the weather is like between where I am and where I'm going, especially if I'm flying across the continent.

Passengers usually don't realize the effort pilots put into a flight. For instance, I try pretty hard to avoid turbulence. I will often call the company dispatcher to see if changing the route of flight might yield smoother air. During the flight, I'll ask air traffic controllers for help in determining if changing altitudes will offer a better ride, or I'll ask them to solicit reports from nearby flights. I want to give my passengers and crew the best ride possible. Turbulence is often unpredictable and sometimes cannot be avoided, but I like the intellectual challenge of finding smooth air.

I've carried about one million passengers so far in my twenty-nine years as a professional airline pilot, and until Flight 1549, not many of them would ever remember me. Passengers may say hello if they meet me as they board, but just as often, they never see my face. After we land safely, they go on with their lives, and I go on with mine.

It's likely that hundreds of thousands of people watched coverage of the Flight 1549 incident, not realizing that they had once placed themselves in my hands for a couple of hours. It's all part of how our society works: We briefly entrust our safety and the safety of our families to strangers, and then never see them again.

I'll often stand at the door to say good-bye to passengers after a flight. I like interacting with them, but you can understand that after all my years of flying, a lot of the passing faces can become a blur. Some passengers stand out—the cranky ones, the first-time fliers who seem so enthralled, the recognizable faces in first class.

One night in the late 1990s, I was flying an MD-80 from New Orleans to New York and the comedienne Ellen DeGeneres was in first class. Shortly after she took her seat in 2D and before we left the gate, my first officer left the cockpit, walked into the front of the cabin, and gave her an enthusiastic greeting. "You are one funny-ass lady!" he told her.

I watched this scene, laughing. I wouldn't have complimented her quite that way, and I'm sure in some HR manual, we're told that we're not supposed to address any passenger as "a funny-ass lady." But Ellen smiled and seemed to take the comment in the right spirit.

We headed back into the cockpit and then flew Ellen, and any other funny-ass passengers on the plane that day, up to New York.

Flights are almost always routine, but every time we push back from the gate, we must be prepared for the unexpected. About a decade ago, I was flying from Philadelphia to West Palm Beach, Florida. At 9 P.M., we were at thirty-five thousand feet, just about fifty miles south of Norfolk, Virginia, when I got word from a flight attendant that a fifty-seven-year-old woman was not feeling well.

From the cockpit, we began the process of getting a radio-to-phone patch to contact a medical advisory service, while flight attendant Linda Lory attended to the woman. Linda got a bit of medical history from the woman's brother and another relative traveling with her, and passed the information up to us in the cockpit. The relatives said the woman had a history of emphysema but hadn't been to a doctor in years.

A few more minutes passed, and as soon as we established communications with the medical service, we got word that the woman was unconscious. Because the aisle was narrow, laying her flat on the floor of the plane was difficult. Passengers nearby were watching it all unfold.

"You have the aircraft," I told the first officer, Rick Pinar. I called air traffic control, declared a passenger medical emergency, and received immediate clearance to a lower altitude and a left turn direct to Norfolk.

"Make an emergency descent and divert to Norfolk," I said to Rick.

What are a pilot's obligations to a sick passenger? We aren't doctors. So how do we determine when a passenger is so ill that an emergency landing is required, diverting the flight to the nearest airport that has appropriate medical facilities, disrupting other passengers' travel plans?

We have access to advice from contract medical services and they and the airline dispatcher help a captain make an informed decision about whether to divert and to what airport. When making such a decision, we have a legal obligation, but more than that, we have a moral obligation to protect life. It's one of the responsibilities we signed up for. It's part of our commitment to safety. If in my judgment I have to land a plane to save a life, I do so.

On this particular flight, we flew as fast toward Norfolk as the airplane could go. There are federal aviation regulations about maximum speeds below ten thousand feet. For jets, it's 250 knots, or about 288 miles an hour. In an attempt to save the woman's life, we went above that speed—over 300 knots. We also made a rapid descent.

Once we touched down, we used heavy braking to shorten our landing roll, allowing us to turn off the runway more quickly. We taxied as fast as was reasonable to the gate.

It was all a bit disconcerting to the passengers. They could see the woman on the floor of the aisle, making no movements. They could feel the heavy braking. They knew we were taxiing faster than usual toward the gate.

Linda, the flight attendant, didn't strap herself into her seat for landing. She was hunched over the woman, trying to save her through mouth-to-mouth resuscitation. It was an heroic attempt on her part.

When we got to the gate, paramedics were waiting for us right on the jetway. They hustled onto the plane as all the passengers watched. They brought a straight-back board, put it underneath the woman, and tried to lift her up. They had trouble turning her on an angle to get her out the door and onto the jetway. It took several minutes to get her off the plane.

I stood on the jetway with the paramedics and the ill woman's relatives. They told me they were on their way to Florida for a funeral of another family member, so an already tragic moment for them was suddenly compounded.

The paramedics worked on the woman on the floor of the jetway for a number of minutes, using drugs, resuscitation equipment, and anything else at their disposal. But it wasn't long before one of them looked up at me and said, "She didn't make it." It's unclear when she died exactly. It may have been while we were taxiing to the gate.

It was a difficult moment, standing there with the woman's family. I tried to say a few consoling words. They weren't weeping; they just looked sad and stricken. My heart went out to them, but I couldn't stay out there for long because I needed to get back on the plane and say something to the passengers.

The passengers had been understanding and cooperative, and had experienced this incident in full view. I felt they deserved to know the truth. And so I got on the public address system.

"The woman who was ill on our flight was under the care of paramedics out on the jetway," I said, "but attempts to revive her were not successful."

There was quiet in the cabin. It was a pretty sobering moment for all of us. Some of the other passengers had watched the woman come onto the plane just like everyone else, put her belongings in the overhead, and settle into her seat. Now, just over an hour after leaving Philadelphia, she was dead.

Because Linda had used emergency medical equipment to help the woman while in flight, we had to wait forty-five minutes for the maintenance staff in Norfolk to replace our medical kit. We also needed to refuel the jet and get a new flight plan. The passengers sat quietly in their seats while we did that.

The woman's family removed their belongings from the plane—they'd be staying with her body in Norfolk—but their checked baggage, and the woman's bags, would have to continue on to Florida with us. There was no time to find their specific bags and remove them from the cargo hold. They'd have to be retagged in Florida and sent back to the family.

About five minutes before we were set to take off again, I called the four flight attendants into the cockpit to join me and Rick, the first officer. As the captain, I was the person ultimately responsible for the decisions made that night. I knew it had been stressful for all of us. I wasn't sure whether the flight attendants felt they could have done more to try to save the woman's life.

First, I thanked them for their efforts. "You did your best. But as tragic as this outcome was, it would be even more tragic if a stressful situation allowed us to be distracted from our duties going forward."

The flight attendants looked a bit ashen and weary. "Rick and I here in the cockpit, we're going to do what we were trained to do," I said. "We'll do our checklist. We'll get the plane into the air. We'll make it safely down to West Palm. I know you have all of your procedures to do, and I know you'll do them as you always have. We'll all need to just fall back on our procedures, and get back into the routine, safe operation that we work so hard to maintain."

The flight attendants headed back into the cabin. We pulled away from the gate with three fewer passengers than had arrived with us.

The flight from Norfolk to West Palm was routine. We arrived just an hour and fifteen minutes late, and I stood outside the cockpit door as all the passengers deplaned.

"Thank you for your patience this evening," I said, nodding at them as they passed. They acknowledged my words with slight smiles or nods of their own. And all of us went to bed that night thinking of the family we had left behind in Norfolk.

Early one Tuesday morning in September 2001, I was driving from my home in Danville to the airport in San Francisco. I had to catch a plane to Pittsburgh, where I was then based, to fly an MD-80 on to Charlotte. I was listening to the radio, an all-news station, and I heard that a plane had just crashed into the North Tower of the World Trade Center in New York.

How could someone be that off course? I thought. *It must have been pretty foggy there.* As I listened to the radio report, I was reminded of the infamous 1945 crash of a B-25 into the Empire State Building, when an Army Air Forces bomber pilot lost his way on a foggy Saturday morning, killing himself and thirteen others. I figured this World Trade Center crash must have been a similar accident.

I parked my car in the airport lot, walked into the terminal, and that's when I heard that another airplane had hit the South Tower and a third plane had hit the Pentagon.

By 6:30 A.M. Pacific time, every airplane in the skies above the United States had been ordered to land, and the FAA had banned takeoffs of all civilian aircraft. It was clear I wouldn't be getting to Pittsburgh that day to fly my scheduled flight. (My particular flight was one of some thirty-five thousand canceled that day nationwide.)

I spent a little time in the US Airways operations office in San Francisco, and there were two crews there. Unlike me, they didn't live in Northern California. They were stranded, and no one knew when planes might fly again. "You'd better get hotel rooms right now," I suggested, "before they're all gone."

I called pilot scheduling and told them that I couldn't make it to Pittsburgh, obviously, and then I went home and watched CNN. As an American and as a pilot, I found the coverage very hard to take. It was so upsetting and disturbing that, at one point, I had to stop watching. I turned off the TV and went into the backyard to compose myself. It was a beautiful day in California, and it was remarkably quiet outdoors. Because all aircraft were grounded, you couldn't hear any airplanes flying anywhere. My ears are always pretty attuned to the sounds of jets, and this saddened me.

On Wednesday and most of Thursday, only the military was flying. I felt anxious about the terrorism and the national ground stop instituted by the FAA, and was eager to return to flying. Like so many pilots, I also felt a renewed sense of patriotism. I wanted to fly to prove our system could function, that we could take passengers safely to where they needed to go, and that the terrorists would not succeed.

On Thursday night, I was able to get on a red-eye to Pittsburgh. On Friday morning, I was set to fly again.

It was pretty chaotic in the crew room underneath the terminal at Pittsburgh International Airport. Not all crew members were able to make it in, and so a captain would say, "I have a first officer but need a flight attendant," and a flight attendant would volunteer to take the trip with him.

Eventually, I was assigned to fly from Pittsburgh to Indianapolis. Not many Americans were yet ready to return to the skies, so we took just seven people to Indianapolis and eight people back from there to Pittsburgh.

There were so few of them, they barely outnumbered the crew. We put them all in first class. Some of the passengers said that they were nervous, and I tried to reassure them with small talk when they boarded.

It was just three days after the attacks, and our planes were still vulnerable to terrorism. But I wanted passengers to know that even though the cockpit doors hadn't yet been strengthened, there was a strengthened resolve among us in the cockpit, and the flight attendants in the cabin. The passengers had strengthened their resolve, too.

"We're determined not to let anything like this happen again," I told a few passengers.

The pilots murdered on September 11, 2001, were the very first victims. And so it was natural for pilots to discuss how we might have responded that day. The reality was that all our training until then had been aimed at preventing or managing a potential highjacking, not a kamikaze-style suicide mission.

For airline employees, life is different now. The airline industry suffered a financial collapse after the attacks, and a great many people at the bottom of the seniority list were laid off. So many of them were good pilots, and they are missed.

The attacks of September 11 don't come into my head as often as they once did. That's true for a lot of Americans. Time has passed. New tragedies have followed. I've piloted hundreds of flights since that day.

But for someone who works for an airline, the reminders are still here, offering reasons for reflection. Sometimes I'll be at Boston Logan International Airport, passing by the gates from which two of the flights departed on September 11—American Airlines Flight 11 from Gate 32 in Terminal B, and United Airlines Flight 175 from Gate 19 in Terminal C.

There are American flags flying outside both of those gates as silent tributes. They are not part of any formal memorial. They were placed there by airport and airline employees. When I pass the flags, I am reminded of the sense of duty I felt on the day of the attacks—to get back in the air, to keep flying passengers to their destinations, to maintain our way of life.

In recent years, I'll often come home from work weary. I've been gone for days. I may have traveled twelve thousand miles. I've endured all sorts of weather or traffic delays. I'm ready for bed. A lot of wives ask, "How was your day at the office?" Their husbands talk about big sales they've made or deals they've closed. I've also had my good days at the office.

One evening I came home and Lorrie was standing in the kitchen. She asked how my day had gone. I began to tell her.

I had piloted an Airbus A321 from Charlotte to San Francisco. It was one of those nights when there wasn't much traffic. Air traffic controllers didn't have to impose many constraints about altitude or speed. It was up to me how I wanted to travel the final 110 miles, and how I would get from thirty-eight thousand feet down to the runway in San Francisco.

It was an incredibly clear and gorgeous night, the air was smooth, and I could see the airport from sixty miles out. I started my descent at just the right distance so that the engines would be near idle thrust almost all the way in, until just prior to landing. If I started down at the right place, I could avoid having to use the speed brakes, which cause a rumbling in the cabin when extended. To get it right, I'd need to perfectly manage the energy of the jet.

"It was a smooth, continuous descent," I told Lorrie, "one gentle, slowly curving arc, with a gradual deceleration of the airplane. The wheels touched the runway softly enough that the spoilers didn't deploy immediately because they didn't recognize that the wheels were on the ground."

Lorrie was touched by my enthusiasm. She noticed that I was telling the story with real emotion. "I'm glad," she said.

"And you know what?" I told her. "I'm guessing no one on the plane even noticed. Maybe some people sensed it was a smooth ride, but I'm sure they didn't think much about it. I was doing it for myself."

Lorrie likes to say that I love "the art of the airplane." She is right about that.

The industry has changed, the job has changed, and I've changed, too. But I still remember the passion that I hoped one day to feel when I was five years old. And on this night, I felt it.

9

SHOWING UP FOR LIFE

In March 1964, when I was thirteen years old, I saw a story on the evening news that I couldn't get out of my head.

My parents, my sister, and I were in our family room, eating dinner on TV trays and watching our black-and-white Emerson TV, a bulky box encased in a blond wood cabinet. As usual, my parents turned the cream-colored plastic channel knob until they came to NBC's *Huntley-Brinkley Report*. David Brinkley was based in Washington, D.C., and Chet Huntley was based in New York, where news had broken about a twenty-eight-year-old woman named Kitty Genovese.

She lived in Queens, and had been stabbed to death outside her apartment. Her neighbors heard her

screams as she was being attacked and sexually as-
saulted by a stranger. Allegedly, they did nothing to
help her.

According to the news report, thirty-eight people
had heard her cries for help and didn't call police because they didn't want to get involved. Their inaction
was later dubbed by sociologists as "the bystander
effect." People are less apt to help in an emergency
when they assume or hope that other bystanders will
step up and intervene.

These initial news reports about the incident would
eventually turn out to be an exaggeration. Some neighbors didn't act because they thought they were witnessing a lovers' quarrel. Others weren't sure what
they were hearing on a cold night with their windows
closed. One person did end up calling the police.

But back in 1964, all I knew was what I was hearing from *The Huntley-Brinkley Report,* and the news
was very shocking to me, and to my family, too.

I found myself thinking a lot about Kitty, and about
her neighbors in New York. What transpired there felt
utterly foreign to me. I couldn't imagine this happening in North Texas. Where I lived, people felt a strong
sense of community while also recognizing that they
would often have to handle their problems and emergencies all on their own. This sense of both fellowship

and self-reliance was necessary in a sparsely populated rural area.

Whatever danger or challenge you faced, you couldn't just dial 911. The nearest police or fire station was too far away. So, at least initially, you would have to deal with it yourself or quickly seek help from your closest neighbor, whose home might be a mile away. By necessity, we had to be self-sufficient. But we also knew that if we needed help, we could turn to our neighbors and they would do their best.

It saddened me to think of these people in New York, in such close proximity to a woman being murdered, and choosing not to help. The police were just a few blocks and an easy phone call away. I couldn't fathom the human values that would allow this to happen. I had never been to New York—in fact, I wouldn't make my first visit there until I was thirty years old—and it was disturbing to me to hear that this could happen in a big city. I talked to my parents about how things seemed so different in New York compared with what we believed and how we lived in North Texas.

I made a pledge to myself, right then at age thirteen, that if I was ever in a situation where someone such as Kitty Genovese needed my help, I would choose to act. I would do whatever I could. No one in danger would be abandoned. As they'd say in the Navy: "Not on my watch."

I now know, of course, that a great many New Yorkers have the same heartfelt urges to help others, and the same sense of empathy, as people anywhere else in the country. We all saw that on September 11, 2001. And I saw it again, firsthand, when Flight 1549 landed in the Hudson, and it felt as if the city rose up at every level to help our passengers and crew.

But back when I was thirteen, and Kitty Genovese was in the news, I felt this real resolve. It wasn't anything I put in writing. It was more of a commitment I made to myself, to live a certain way.

I'd like to think I've done that.

I've come to believe that every encounter with another person is an opportunity for good or for ill. And so I've tried to make my interactions with people as positive and respectful as I can. In little ways, I've tried to be helpful to others. And I've tried to instill in my daughters the notion that all of us have a duty to value life, because it is so fleeting and precious.

Through the media, we all have heard about ordinary people who find themselves in extraordinary situations. They act courageously or responsibly, and their efforts are described as if they opted to act that way on the spur of the moment. We've all read the stories: the man who jumps onto a subway track to save a stranger, the firefighter who enters a burning building knowing the great risks, the teacher who dies protecting his students during a school shooting.

I believe many people in those situations actually have made decisions years before. Somewhere along the line, they came to define the sort of person they wanted to be, and then they conducted their lives accordingly. They had told themselves they would not be passive observers. If called upon to respond in some courageous or selfless way, they would do so.

Lorrie and I have done our share of very small things to help the greater good. A year ago, we were stopped at a red light in our hometown of Danville and we saw a female pedestrian in her late forties walking her small dog across the street. Lorrie saw the driver in front of us about to make a left turn. "He's going to hit her!" Lorrie screamed. "He's going to hit her!" And he did.

It was unclear to us whether the driver of the car was not paying attention or if the sun was in his eyes—but the woman was knocked unconscious, and her dog ran loose. She was lying facedown in the street and I was one of the first people to get to her.

I made sure someone called 911 and that someone checked that she had a pulse and was breathing and not bleeding, while I helped direct traffic around her before the police arrived. I was impressed with the other motorists. They recognized the gravity of the situation and were patient. No one was honking. No one tried to pull out and drive around the scene. It seemed as if everyone had the right attitude, the right values, and did the right thing. Someone got the woman's dog. Another person found the woman's cell phone and pulled up her daughter's phone number from the phone's contact list. The woman was taken away in an ambulance and survived.

I was pleased to see the people of Danville respond so well, and I was glad to be involved.

I've been moved and impressed by my daughters' eagerness to help others.

Kate raised and trained two puppies for Guide Dogs for the Blind. The program sent us our first puppy, a yellow Labrador retriever named Misty, in November 2002. Kate immediately fell in love with the puppy. She worked day after day helping Misty understand verbal orders. To get a puppy to relieve herself on command, the trainer has to wait for her to go to the bathroom, and then say the command "Do your business!" The idea was that Misty would then associate the words with doing her business, and when serving a person with disabilities, would be able to "relieve on command."

Kate, then nine years old, took her responsibilities very seriously. One stormy day, I looked out the window and saw she was outside in the pouring rain, wearing her yellow slicker and galoshes, waiting for Misty to relieve herself so she could tell her, "Do your business!"

I called Lorrie over to the window to watch. We were proud of Kate. She was so responsible. And she loved that dog so much.

Once Misty was trained, we had to give her back to the organization so she could be placed in a home with a person who needed a guide dog. We knew that the good-bye would be very hard on Kate. "Recall Day" turned out to be Valentine's Day 2004, when Misty was fifteen months old. Kate held herself together until it was time to leave Misty behind. Then she began bawling. For a while after that, she said she didn't want to allow herself to fall in love with anything or anyone because it was going to be too hard when it was over. She said losing Misty was the first time she'd ever had her heart broken.

Through it all, though, she saw the great value of the guide dogs program. "We're helping people," she'd say, "and giving them their freedom back. It feels good to be able to do that. Besides, it's fun to have a puppy."

Kelly, meanwhile, is one of the most empathetic people I know. Starting in preschool, she always has been the kid who'd raise her hand and volunteer to be the teacher's helper. She also embraced "Books for the Barrios," the brainchild of the wife of a former naval officer and American Airlines pilot. The program has sent twelve million books to impoverished students overseas.

In second grade, Kelly's class took a field trip to the organization's warehouse in Concord, California. They learned about all of the disadvantaged kids on the outlying islands in the Philippines. They were told that many of the children slept on dirt floors, and welcomed the cardboard boxes that Books for the Barrios were packed in. Families broke down the boxes and used the cardboard as mats to sleep on.

Kelly was moved by what she heard on that field trip, and for her eighth birthday party, she decided on her own to ask her friends to bring books and gifts for children in the barrios. The children were instructed, when selecting gifts for Kelly, to pick presents that were appropriate for children in the Philippines. The party was held at the warehouse, and Kelly placed the wrapped gifts into shipping boxes. She and her friends then spent an hour helping pack donated books into boxes they decorated.

Everyone's reputation is made on a daily basis. There are little incremental things—worthwhile efforts, moments you were helpful to others—and after a lifetime, they can add up to something. You can feel as if you lived and it mattered.

Until Flight 1549, I had assumed that I would always live a pretty anonymous life. I'd try to do my job to the best of my ability. Lorrie and I would try to raise the girls with the values we cherish. I'd make an effort to volunteer for worthy projects. Perhaps, I thought, at the end of my life, in aggregate, it would all add up to my being able to say I'd made a difference to others and to my community in some small way.

Actually, I live in several communities. One is Danville, of course. But another is the community that keeps re-creating itself in the nation's airports. It's a community of familiar faces—airport workers, my colleagues at US Airways, the crews from other airlines—that also includes thousands of strangers who repopulate the terminals every day.

An airport is not always an easy place to connect meaningfully with other people. We're all coming and going, trying to get somewhere else and then home.

But there are little ways to show humanity, and I've admired those who find ways to do so.

A pilot's job, first and foremost, is to fly the airplane safely, delivering passengers from Point A to Point B. We have checklists outlining a host of other tasks, too. But there are many things that are not in our job description, things that are the responsibility of gate agents, baggage handlers, skycaps, caterers, cleaners.

Most of these people do their jobs well, but an airport and an airline are not perfect systems. That can be frustrating for travelers and for those of us in the industry. If I can help things along, I try to do so.

There was one time when we had flown from Philadelphia to Hartford, Connecticut, landing at 10:30 P.M. A young couple in their thirties with a toddler waited and waited on the jetway for their stroller, but it never showed up. I wanted to help them. My attitude with passengers in these situations is this: I've gotten you this far. I'm not going to leave you hanging now.

I went down the stairs and out to the ramp and talked to the baggage handlers. Then I came back and told the couple that the stroller was either lost or left in Philly. "Come with me," I told them.

I walked the couple to baggage claim and showed them where to file a claim. It was late. The lights in the terminal were being shut off. If I didn't get them to the right place, they'd be stuck in the airport with everything closed, including the baggage office.

A flight attendant saw me helping them and commented that not every pilot or flight attendant would bother to help. It was an awfully simple thing I had done. I barely had to walk out of my way, since I was headed to a hotel van right outside of baggage claim.

And yet, I understood completely what this flight attendant meant.

A lot of people in the airline industry, and especially at my airline, US Airways, feel beaten down by circumstance. We've been hit by an economic tsunami. Some people feel their companies have held a gun to their heads, demanding concessions. We've been through pay cuts, givebacks, downsizing, layoffs. We're the working wounded.

People get tired of constantly fighting the same battles over and over again every day. The gate agent hasn't pulled the jetway up to the plane in time. The skycap is supposed to bring the wheelchair and hasn't. (I've helped more than a few older people into wheelchairs and pushed them into the terminal myself.) The caterer hasn't brought all the first-class meals. Catering companies always seem to be the lowest bidders with the highest employee turnover. At the end of a long day, you and your crew will get off the plane and make your way out of the terminal, but the hotel van isn't there when it's supposed to be.

All of this stuff beats you down. You get tired of constantly trying to correct what you corrected yesterday.

Many pilots and other airline workers feel that if they keep picking up all the slack, those who run the companies we work for will never staff the airlines properly, or do the training necessary, or hire the contractor who will be most responsible about bringing wheelchairs. And my colleagues are right about that. In the cultures of some companies, management depends heavily on the innate goodness and professionalism of its employees to constantly compensate for systemic deficiencies, chronic understaffing, and substandard subcontractors.

At all airlines, there are many employees, including in management, who care deeply and try to make things better. But at some point, it can feel like a fine line between letting passengers fend for themselves and enabling the airline's inadequacies. And so it becomes a decision whether to do the simple, easy act of walking a young couple and their toddler to baggage claim.

My way of handling these issues is to fight to improve the system but still help those I can.

There was another incident late one night at the airport in Charlotte. We were delayed because of weather and air traffic issues, and as my crew stood on the curb waiting for the hotel van, a woman saw me in my pilot's uniform and approached me. She was around fifty years old with short brown hair. She had no purse, no luggage, only a cigarette in her hands.

She said she and her family had flown in on US Airways, and she was changing planes in Charlotte, on her way to another city. Her family was back at the gate, where their plane was delayed because of the weather.

"I asked an airport employee where I could smoke a cigarette, and he sent me out here to the curb," she told me. But without thinking, she had left her purse and boarding pass with her family at the gate, on the other side of security. And worse, a few minutes earlier, at 10:30 P.M., the security checkpoints had closed. The Transportation Security Administration is a bureaucracy. When it closes, it closes. At 10:30 P.M., you can go through. At 10:31 P.M., you can't. So she was stuck.

I could have told her that I was unable to help her, then gotten into the hotel van and driven off. But that wouldn't feel right. I took out my cell phone and called a couple of people in operations. I gave them her name, her cell-phone number, and tried to see if they could somehow help her get back to the gate—or at least get her a voucher for a hotel room.

I don't know what became of that woman that night. But I felt I had to try to help her. As a human being, I couldn't just go to the hotel and leave her behind.

Again, it hardly took any effort on my part. Besides, I don't want to go through life as a bystander.

When there are maintenance issues or other delays, I believe in telling passengers exactly what is going on. Sometimes a plane has to be taken out of service after passengers are already loaded and ready to go. I don't like to leave it to flight attendants to give the bad news. I get on the public address system, and offer up the details. I have stood in the front of the cabin, where the passengers can all see me, and I've said: "Ladies and gentlemen, this is the captain. This airplane has to be taken out of service, so we're going to have to change airplanes. We'll need to get off this plane and the gate agent will send you to the new gate. I appreciate your patience, and I apologize for the inconvenience."

When I do this I also want to protect the flight attendants from any kind of whining or abuse as people deal with the delay. "I'm the one responsible for this change," I'll say. I'll stand at the door as each passenger deplanes, looking them all in the eye and nodding. I want them to know that if they have an issue, they should talk to me, not take it out on the crew.

I've learned that word choice is so important. When there's a delay, I like to address passengers by saying: "I promise to tell you everything I know as soon as I know it." I've found such language makes a world of difference. It's inclusive. It tells passengers our intention is to give them the whole truth, and it lets them know we trust and respect them enough to share this truth. Not being honest up front might avoid hard questions early on, but then there can be consequences for the flight attendants later, when they have to deal with passengers who feel they were lied to. It also hurts the reputation of the airline.

If passengers decide they haven't been dealt with honestly, they get on their connecting flight feeling angry. Then a vicious cycle sets in. Passengers have already formed a negative impression of the airline, and through the filter of that negativity, they start finding things that support their preconceived notions. They discount things that are positive as being due to chance, and they view negative things as supporting their belief that "this is a lousy airline."

I can avoid all that just by being straightforward with passengers from the cockpit.

For the most part, I find passengers to be considerate and understanding. Flying is not the genteel activity it once was, but given that passengers are all cooped up in a relatively small space, and that can be aggravating and uncomfortable, they tend to rise to most occasions.

A lot of times I feel for passengers, and for the situations they find themselves in given all the issues that define air travel today: enhanced security checks, more-crowded cabins, long flights without food service. I'll try to do what I can.

Passengers often don't know when efforts are made on their behalf by the crews on airplanes. Sometimes, we're pulling for them—quietly or under the radar.

For instance, the airlines want flights taking off on time. It makes your airline look better when your on-time rate is higher than other airlines'. Gate agents are judged on their ability to deliver on-time departures. This can make for tension among airline employees, and it's certainly not always best for passengers.

And so sometimes, I've felt obliged to stand my ground.

There was one Sunday afternoon when I was flying from West Palm Beach to Pittsburgh. There was a fairly substantial standby list of people hoping to get on the plane. Everyone with an assigned seat was loaded on, and then the gate agent came on the plane to say that he would close the door. He wanted us pushing back on time. I told him there were still two empty seats.

"Whoever is next on the standby list, why don't you send them down?" I said.

The agent was having none of it. He wanted us closing the door and pushing back. He knew that his station manager's job-performance evaluation is based partially on statistics for on-time departures. He didn't want to get any grief from his superior, and so he didn't want to take a few more minutes to get two more passengers on the plane.

I understand the ramifications for everyone in the airline system. The station managers dump on the agents. The agents push the crews to load faster. The statistics-driven system is not forgiving if, say, six people in wheelchairs have to be loaded, and that slows down boarding.

Anyway, this gate agent and I were at odds over these two empty seats that I wanted to fill. I had to speak up.

"Let's remember why we're here," I told him. "We're here to get paying customers to their destinations. You have two paying customers out there who want to be on this plane, and there are seats available for them. So I say, let's quickly get them on board."

I prevailed. After all, the policy manual says the captain is in charge. And so the two passengers at the top of the standby list were invited onto the plane, and we ended up pushing back two minutes late. We may have been a minute or two late to Pittsburgh.

The following Tuesday was my day off, and my phone rang at home. It was the assistant chief pilot. He told me that he had a letter from a passenger service supervisor in West Palm.

"They say you interfered with the boarding process, delaying the flight," he told me. And then he started reading me the riot act. He talked to me like a disciplinarian, as if I were some renegade cowboy in the cockpit, keeping the gate agents from doing their jobs.

I was a bit peeved by this phone call.

"I care deeply about doing a good job," I told him, "and I think there are two possibilities regarding this incident. The first possibility is that the agents were following company procedures, and the company procedures are flawed. The second possibility is they weren't following procedures, in which case they should. We had one hundred and fifty seats, two of which were empty. I wanted to see them filled. I think that's good for the company and good for the passengers."

The assistant chief pilot didn't seem pleased that I was pushing back. But we let the matter be.

Six months later, on another Sunday, I found myself in the same situation. Empty seats. People in the boarding area eager to take them. The agents wanted us to close the cabin door, I insisted that we load the passengers, and our flight left the gate six minutes late.

The agents wrote me up again. And the assistant chief pilot called me again. He was in a pissier mood this time. "The chief pilot wants to give you two weeks off without pay," he told me.

My union rep ended up talking to management and they never went through with their suspension threat. After all, I wasn't alone. Many captains were having to fight this battle repeatedly. And then one day, a few months later, management came out with a new memo. It stated that passengers are not to be left behind if seats are available to them. I smiled when I read that.

All of us have little battles we can choose to take on or to skip. Some captains feel as I do about these sorts of things, and they fight. Others acquiesce and give up. None of us likes leaving passengers at the gate, but some have decided: "I can't fight so many battles every day."

I guess I haven't had what I call "a sense of caring" beaten out of me yet. I empathized with those standby passengers. But as important, leaving them behind just would have felt wrong. And so I acted.

These are minor things, I know. But I feel better about myself when I make these kinds of efforts. And it's nice to feel I'm doing a little good in the process.

. . .

I've read a great deal as I've commuted from San Francisco to my base in Charlotte. The trip across the country seems to go faster when I'm engrossed in a book. My tastes haven't changed much since I was a boy: I continue to be drawn to history.

I have read a few terrific books about the nation's Medal of Honor recipients. Each of their stories is inspiring. But I remain particularly haunted by the story of twenty-three-year-old Henry Erwin, a U.S. Army Air Forces radio operator from Alabama whose heroism during World War II was astounding. On April 12, 1945, Staff Sergeant Erwin was on a B-29 mission to attack a gasoline plant in Koriyama, Japan. One of his tasks was to help the bombers see their aim points by dropping a phosphorus flare through a tube in the floor of the B-29. the device exploded in the tube, and the phosphorus was ignited, blinding Erwin and engulfing him in flames. Smoke filled the airplane. Erwin knew the flare would soon burn through the floor, igniting the bombs in the bomb bay below, destroying the B-29 and probably killing the crew.

Though Erwin was in excruciating pain, he crawled along the floor, found the burning flare, and held it against his chest with his bare hands. He brought it up to the cockpit, screamed to the copilot to open his window, and heaved it out, saving the other eleven men on board.

Erwin was expected to die within days from his injuries, and the decision was made by General Curtis LeMay to award him the Medal of Honor before he succumbed. The problem was, there was no Medal of Honor to be found in the Western Pacific. The closest one was hours away in a glass display case in Honolulu. And so an airman was dispatched in the middle of the night to go pick it up. When he couldn't find the key to open the display case, he broke the glass. He collected the medal, and put it on a plane bound for Guam, where it was pinned on the still-alive-and-conscious Staff Sergeant Erwin, wrapped head to toe in bandages.

Erwin surprised everyone, living through forty-three operations. He remained hospitalized until 1947, and after he was released, his burns left him scarred and disfigured for life. Yet he continued to serve his country as a counselor at a Veterans Hospital in Alabama. He died in 2002.

Who among us could have brought ourselves to lift that white-hot flare to our chest with our bare hands? Presented with that situation, I assume I would have let it burn through the floor of the B-29.

Knowing that there have been people like Erwin, capable of doing such extraordinary things—acts that are truly beyond comprehension—I feel that the least I can do is be of service in whatever very small ways are available to me.

Sometimes that means recalling how I felt as a thirteen-year-old, when I first heard the story of Kitty Genovese, and made a vow about the kind of person I hoped to be. And sometimes it means attempting the smallest of acts—helping a couple find a lost stroller, or enabling a standby passenger to get the last seat on a departing plane.

10

ANYTHING IS POSSIBLE

Lorrie and I have a favorite hill, and we're very lucky, because it is within minutes of our house, on a large piece of open land right at the edge of our neighborhood in Danville. We hike up there together to think, to breathe, and to appreciate. It's a pretty magical place.

Most of the year, everywhere you look on that hill, there are acres of tall native grasses, in every variation of brown and gold. Later in the spring, for a short while, the grass turns green and more lush. Brown or green doesn't matter to me. I appreciate the beauty there in every season.

On the afternoon of January 11, 2009, Lorrie suggested that we take a walk up that hill. It was a Sunday, and I was scheduled to leave early the following morning for the trip that would end, four days later, with Flight 1549.

We had a lot on our minds that day. Like so many Americans, we were worried about the economy, and how our serious financial issues might be resolved. I continued to be very concerned about the Jiffy Lube franchise that hadn't renewed its lease on our property, and about our ability to keep up the mortgage payments on it. It was a problem with no easy solution. My focus can narrow when thinking about our personal problems or the economic woes of the airline industry, and Lorrie has a wonderful ability to help me change my perspective.

We were sitting in the kitchen, and Lorrie knew what might help. "Come on," she said to me, "let's go for a walk."

And so we hiked up the fire road that narrows into a trail on that steep, beautiful unnamed hill. We stopped at the top to look into the valley below. It's a gorgeous panorama of neighborhoods in one direction, and pristine open spaces in the other. The sights from that hill can literally widen your view of the world. Somehow, your troubles get put into perspective. The view restores and renews you.

On that day, Lorrie and I were quiet for a little while, just taking it all in, and then I said to her: "Looking out there makes you feel like anything is possible."

She smiled at me. She already knew it, without me having to say the words. That's Lorrie. If you want to discover the benefits of believing that *anything* is possible, hike up a hill with her. You'll be inspired and reassured.

Lorrie is an exceptionally strong woman, and as I have watched her grapple with various issues in her life, and the challenges in our own family, I've learned a great deal about the power of optimism and acceptance, and about the responsibilities all of us have to carve a path to our own happiness.

She and I are a bit different. I'm a believer in "realistic optimism," which I consider a leader's most effective tool. That's short-term realism combined with long-term optimism. Lorrie understands the value in that, to be sure, but she also sees how embracing full-on optimism about life's possibilities is good for your health, your relationships, and your sanity.

Lorrie speaks frankly and from the heart, and she's able to take her life and pull from it moments and experiences that resonate deeply with other women, literally changing their lives. That's what she does in her career as an outdoors fitness instructor, heading a one-woman operation she calls "Fit and Fabulous . . . Outdoors!" She takes groups of women on long hikes. They'll go up one side of a mountain, and by the time

Lorrie brings them back down the other side, they aren't the same women anymore. They've seen the world and themselves in a new way. Sometimes I'll drive the women to the trailhead or back home from it. I've waited for Lorrie at the bottom of a mountain when she and the women in her groups have returned. It's remarkable to watch.

Granted, I'm Lorrie's husband and I love her, so this may sound overstated. But those who've walked up a mountain with her know just what I'm talking about.

One of Lorrie's friends, Helen Ott, who has joined her on numerous fitness hikes, puts it this way: "Lorrie is like a bright light." Helen talks about all the fun she has on these walks, because Lorrie is such a good story-teller and is so supportive of other women. "She makes people feel confident in their abilities," Helen says, "and she makes them feel good about themselves."

Lorrie's embrace of exercise—and the idea that it is best done with others, and in the great outdoors—was actually a journey that began very uneasily for her. She speaks openly with women about how she was "the quintessential chubby girl" for most of her childhood. She has worked to understand the impact her dad's alcoholism had on her eating habits and on her sense of herself growing up. Hers was not a painless childhood, but she isn't one to make excuses.

Lorrie was overweight as an adult, too, and that was exacerbated by the fertility drugs she took trying to get pregnant. The drugs left her thirty-five pounds heavier, and feeling deeply wounded. Unable to conceive a child, she felt that her body had betrayed her. Even after we adopted Kate and Kelly, and even though we felt our family was complete and perfect, her feelings remained raw.

"I fell madly in love with the girls the moment we brought them home in our arms," Lorrie has explained to her clients. "Sully and I felt as if we had won the baby lottery. But those feelings of betrayal, they didn't magically go away. When the infertility ordeal was over, I had two incredibly beautiful daughters that I loved with every fiber of my being, but I was angry at my body."

A decade ago, just before Lorrie turned forty, she decided that she would try to let go of the anger and make peace with her body. First, she joined a gym. But walking on a treadmill and going nowhere seemed unsatisfying. She was still having what she called "negative conversations" with her body parts. She told me she felt "awkward" at the gym. "The more I focused on my butt, the bigger it seemed to get," she'd say. Like a lot of people, she was trying to lose weight by beating her body into submission.

The mind-body connection is powerful, of course, and the fact that she didn't like the body carrying her through life was a big issue. Then she took a class at a local gym with a woman named Denise Hatch, who put things in perspective: "Be grateful for what your body can do, rather than focusing on what it can't do. You can't have children. That's hard on you, I know. But your arms and legs work. You're healthy. You have two daughters who need to see you modeling healthy behavior. So all of this negative body image talk and thoughts have to stop right now."

Lorrie learned that it was vital to find a way of exercising that she liked. "If you're not a runner, then be a walker, a hiker, a dancer," she tells women now. "Just be brave. Find your thing and do it. As with everything in life, if you like doing something, you will do it more often."

In Lorrie's case, hiking liberated her. Walking outdoors and seeing a red-tailed hawk gliding overhead or looking out over the carpetlike hills of California or feeling the softness in the summer wind—she realized she was having spiritual experiences she'd never find on a treadmill. And her enthusiasm was contagious. She wanted to hike every day, and to take me and the girls with her.

"I think I'm in love," she told me one day, ". . . with exercise."

Lorrie would go on to be a fitness expert on the San Francisco ABC-TV affiliate, hosting regular segments about how women can incorporate the outdoors in their quest for better health. And she takes groups of women on regular hikes, listening to the stories of their lives as they walk, and sharing her own.

"The body that betrayed me for so long responded to the outdoors," she explains to them. "Exercise gave me the confidence that had eluded me. It made me a better mother, wife, and friend. And I hiked off those thirty-five extra pounds."

Lorrie is frank. "As women, we have to become comfortable with our bodies. That's crucial. A woman who isn't comfortable will turn off the lights at night and say to her husband, 'Please don't touch me.' When a woman is happy in her own skin, she's more willing to let her partner be close."

For years now, Lorrie has included me as a character in her repertoire of inspirational stories. I'm not sure I want to know everything discussed high in those mountains about our private lives. But I'm happy with Lorrie's basic message: "Hiking," she says, "has reinvigorated my marriage."

It was Lorrie's idea. She wanted us to hike together to the top of California's Mount Whitney, which is in the Sierra Nevada range, southeast of where we are in Northern California. At 14,505 feet, it's the highest peak in the contiguous United States.

This was fairly early in Lorrie's discovery of hiking, and she arranged for eight couples to go together. She got the necessary U.S. Forest Service hiking permit, but one by one, for scheduling reasons or because they hadn't trained well enough, each of the other couples dropped out. The sixteen-person hike became a two-person hike—me and Lorrie—but we decided, what the heck, we'd still do it.

We trained for the adventure faithfully. Whenever I was home from a trip, we'd put on our running shoes and run over to a shopping center a mile from our house, where there is a series of stairs leading up a hill to a parking lot. We'd run up and down the stairs fifteen or twenty times, and then we'd jog home.

We kept going to the gym to lift weights, and we went on practice hikes locally, carrying weights in our backpacks. We also did a lot of biking up Mount Diablo, just northeast of Danville.

Lorrie believes that to meet your goals in life, it's important to write them down. But that's not enough. You also need to take what she and others call "authentic action" every day to achieve them. That means you have to knock on a door, or make a phone call, or do something concrete to get you closer to your goal. When training to hike the tallest mountain in the continental United States, you have to get out every day and prepare. She made sure we did that. In the middle of our training, I hit a patch of gravel while riding a mountain bike on Mount Diablo, breaking my pelvis. I was out of work for six weeks, and it made getting back to preparations for Mount Whitney that much more challenging.

Lorrie felt that, not unlike our adoption journey, training for the hike would be good for us as a couple. We needed each other for emotional support. When one of us was tired, the other would offer encouragement. And these moments of rallying for each other would be good practice for the support we'd have to give each other on the actual hike.

Our ascent of Mount Whitney was set for September 2, 1999. We got a babysitter for the girls, and rather than driving the seven hours southeast from our house to the mountain, we decided to rent a Cessna Turbo 182RG (a four-seat, single-engine plane) and fly there. It was pretty romantic, just the two of us, heading off to test ourselves in the wilderness.

We planned to complete the hike in one day, but that meant we'd have to start very early. We stayed in a motel near the mountain, woke up at 3 A.M., and were on the trail at four-fifteen, wearing our headlamps and backpacks, ready to go. The trailhead starts at 8,300 feet, and if we could make it to the top and back, it would be twenty-one miles round-trip.

In our backpacks we had rain gear, hats, gloves, spare batteries, matches, power bars, water, peanut-butter sandwiches, and other essentials. I also had brought along a gallon-size plastic bag with my mother's ashes. She had died the January before, and I thought the mountain might be an appropriate place to spread her ashes.

My dad had passed away four years earlier, and after living a pretty traditional life with him, my mom had really come into her own in her final years. My father had been more of a homebody, and my mom had loyally stayed on the home front with him. But once he was gone, she did a great deal of traveling with friends. It was as if she was making up for lost time. She embraced every part of living she could, and it was wonderful to see that. Lorrie and I thought it would be fitting to bring her ashes to this tallest peak so we could set her free in the wind, to continue her travels.

We started our hike well before sunrise, but the moon was half full, and straight up in the sky. There was so much light from the moon that our bobbing headlamps were almost unnecessary.

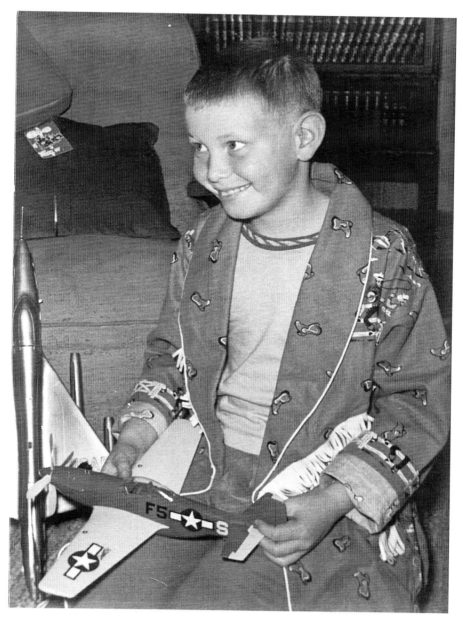

I was lucky to find my life's passion at a very young age. I have a clear recollection that at age five I already knew I was going to spend my life flying airplanes. Here I'm about eight years old and thrilled to have received a model airplane from my parents on Christmas morning.
(Author's Collection)

My mother, sister, and me in our Sunday best, Easter 1955.
(Author's Collection)

My parents on their wedding day, April 1948. *(Author's Collection)*

Growing up in Denison, Texas, was a wonderful experience. My family lived far enough outside of town that there was a multitude of opportunities for a young boy to find adventure, explore the world, and develop some independence. I have fond memories of taking the boat out on Lake Texoma. Here I am with my father and sister in the summer of 1960. *(Author's Collection)*

This photograph was taken in late 1968, shortly after I received my private pilot certificate under Mr. Cook's careful tutelage. Here we are commemorating my mother and sister's first flight with me as their pilot. *(Author's Collection)*

I graduated from Denison High School in May 1969. Following the ceremony, I took my grandparents for their very first flight on an airplane. They were leaving the following day for a trip to Rome and wanted to make sure they were prepared for their TWA flight. (*Author's Collection*)

My father in his naval officer's uniform, 1942. He grew up during the Great Depression of the 1930s and served his country in World War II. He was a member of the Greatest Generation, and his values still inspire me: a sense of civic duty, service before self, and a willingness to share sacrifices.
(*Author's Collection*)

My first flight in a military jet was during my freshman year at the United States Air Force Academy. It was an orientation aboard a Lockheed T-33, a flight designed to remind us of the light at the end of the tunnel, and the first moment when I knew that I was well on my way to realizing my dream. *(Author's Collection)*

During the summer of 1971, I was assigned to Bergstrom Air Force Base in Texas. This was a routine training flight in the back seat of an RF-4C. *(Author's Collection)*

One of my first assignments in the United States Air Force was flying fighters at Luke Air Force Base near Glendale, Arizona. Here I'm about to take a training flight in 1975 with Dave, my Weapons Systems Officer, in the F-4 Phantom II. *(Author's Collection)*

On June 6, 1973, I graduated from the United States Air Force Academy,
receiving my diploma from Superintendent Lt. Gen. Albert P. Clark.
Upon graduation, I was named "Outstanding Cadet in Airmanship" in
recognition of all I'd learned flying airplanes and gliders and parachuting.
(Author's Collection)

Lorrie and me on our wedding day, June 17, 1989.
(Author's Collection)

Top: We took this photograph on the eve of our hike up Mt. Whitney. Despite the enormity of what we were about to do, we knew that we were prepared and ready for the challenge. (*Author's Collection*)

Bottom: To train for the Mt. Whitney hike, Lorrie and I took the girls on a scouting trip in August 1999. We flew up from Livermore to Bishop in a Cessna T182RG. This is Kelly in the back seat, ready for takeoff. (*Author's Collection*)

Lorrie and I took the girls to Washington, D.C., for their spring break in 2002, and they humored me with a trip to the Smithsonian Air and Space Museum. *(Author's Collection)*

Lorrie's love of the outdoors is infectious, and we've shared many adventures, including a snowshoeing expedition in Yosemite in early 2000. *(Author's Collection)*

My mother taught first grade in Denison for more than twenty-five years. When I returned recently for my fortieth high school reunion, I was struck by how many people had been in one of her classes and wanted to share their fond memories of her. She touched so many young lives throughout her career. (*Author's Collection*)

My mother was an excellent musician and had a deep appreciation for the classics. Here we are in Denison, shortly after my father's death, sitting at the piano as she's passing the joy of music on to my daughter Kate. (*Author's Collection*)

Top: In July 2001, Kate and Kelly first rode on a commercial flight that I piloted. While I'm sure the prospect of Disney World was the most compelling part of the trip for them, it was a thrill for me to be able to show them what I do for a living and expose them to my love of flying. (*Author's Collection*)

Left: Before airport security got to be what it is today, Lorrie sometimes brought the girls to the airport to say good-bye before I departed. This was taken in the summer of 1994 when I was flying the Boeing 737, and Kate came to see me off to work. (*Author's Collection*)

My first flight on the Airbus, August 2002. (*Author's Collection*)

On Memorial Day in 2001, we talked about the importance of the holiday with the girls, and Lorrie wondered if my Air Force uniform would still fit. Remarkably it did, though it was a little tighter in a few places. It was a wonderful opportunity to teach Kate and Kelly about the men and women who have made tremendous sacrifices in serving our country. (*Author's Collection*)

Lorrie and I have volunteered with Guide Dogs for the Blind for seventeen years, and our breeding dog, Twinkle, has given birth to four litters over that time. My daughter Kate has trained two puppies, and I am so proud to watch both her and Kelly work with the dogs to make sure they are ready for their future owners. (*Author's Collection*)

Lorrie and I have been through a lot together. She's an exceptional woman, and I am grateful for all of the support and joy she brings to our family. (*Nigel Parry/CPi Syndication*)

The predawn darkness was magnificent. Astronomers would say "the seeing was good." The air was stable, and so the stars were bright and clear, without much twinkling. It was almost as if we could reach out and touch them.

At first, we were walking in the shadows of tall trees, wearing just light jackets. Once the sun started rising and warmed the mountain, we were able to put the jackets into our backpacks.

The sunrise was spectacular. We were hiking on the eastern side of the mountain, facing west, and one peak behind us was perfectly aligned with the sun, forming a triangularly shaped shadow on the expanse of Whitney ahead of us. As the sun got higher, the black triangle moved down the face of the mountain. It was an amazing sight.

We were also fascinated by how the mountain changed as we climbed. With each change in elevation, we traversed different zones with varying terrain and plants. We encountered marshy areas and some lakes and streams, but as we got higher, the vegetation became more sparse. Portions of the trail were rugged and rocky, and at one point we had to scramble over large boulders. Then the altitude began taking its toll on us. We knew this would happen—we had read the books—but that made it only a little easier to handle.

Lorrie had a raging headache, and both of us got slug-gish and very tired.

We kept reassuring each other with an old line that marathon runners use: "It's not twenty-six miles. It's one mile, twenty-six times."

We had another mantra: "Anyone can hike Mount Whitney. You just point your feet in an uphill direc-tion, and put one foot in front of the other." We kept repeating that.

We lost our appetite, which is also common. We knew we had to force ourselves to eat, because we'd need our energy. The guidebooks had told us to bring our favorite foods, even junk food, because we'd be more apt to eat something we liked. It was remarkable to see what happened every time we pulled something to eat from our backpacks. Blue jays would try to land on our shoulders or backpacks to take the food away. Large ground squirrels called marmots would come out of the rocks, almost out of nowhere, and would also try to grab their share. They were all obviously very used to humans and knew that where there were people, there was food.

At thirteen thousand feet, the narrow trail crossed over the top of the mountain and there was a sheer drop-off. We were well above the tree line at this point, and it looked as barren as the surface of the moon.

Lorrie got teary, in part from exhaustion and also, she admitted, out of fear. It was pretty intimidating looking down. She wondered if we really needed to reach the exact summit to release my mother's ashes.

"Why don't we just let your mother out here?" she asked. "Your mom would understand. I know she would."

I wanted to keep going. "We can do it," I told her. She smiled weakly at me, and we pressed on.

By one-fifteen, we were within sight of the summit—maybe an hour from reaching it. But hours earlier, when we began the hike, we had established a turnaround time of one P.M. We knew we needed enough energy and daylight to make our descent, and we didn't want to take any risks that would hamper our ability to return safely. Part of us wanted to continue on. But we deferred to good judgment. We resisted temptation and made a smart decision: We had come far enough.

I was understandably emotional as I reached into my backpack and took out my mom's ashes. I opened the bag, and it was a powerful moment when I let go of her ashes, and watched them take off so easily into the wind. It was a clear blue day, not a cloud in the sky, and the ashes fluttered into the breeze and just kept going.

"I hope she enjoys her travels," Lorrie said, and I wasn't able to say much in response. I just watched.

Once that simple ceremony was over, Lorrie and I allowed ourselves to appreciate the majesty of the view. "Our worries seem pretty small in comparison to all of this, don't they?" Lorrie said to me. "It puts life in perspective."

We rested for a bit, taking it all in. But we couldn't stay there too long. Our hike was only half over at that point.

Descending the mountain was almost harder than the hike up, because we were so drained emotionally and physically. By the end of a hike like this, every part of your body that could possibly chafe against another part of your body has done so.

When we reached the bottom of the trail at 8:15 P.M., again in darkness, we felt absolutely exhilarated despite our exhaustion. We were immensely proud of ourselves. Lorrie, who had spent years believing her body had let her down, recognized that in so many ways, her body had come through for her.

Flying home the next day in the rented plane, I circled over the mountain a few times, and we looked down at it with awe. We both joked that it was a good thing we hadn't flown over it on the way there, because from the air it looked too formidable and steep.

"Wow," Lorrie said to me. "Can you believe we did it?"

On the plane, as we headed over the mountain and then northwestward toward home, Lorrie, inspired, took out a pen and wrote a "gratitude letter."

She wrote of how the mountain helped bring clarity to her life: "I realized how small our daily 'stuff' is. The mountain was here long before we were, and will be here long after we are gone. The fabric on my family room chairs really seemed insignificant by comparison. But what seemed supremely important during the hike was the giggling and laughter of Katie and Kelly, even when we want it quiet, and the love of our family—those living and those who have left us."

Lorrie is well on her way to having led "a well-lived life." She makes her way with passion and purpose, and by doing so reminds others of what is possible. I'm grateful to have shared the trail for so much of it.

Lorrie is always on the lookout for inspiration, and a couple of years ago, she heard Maria Shriver speak at her annual California Governor and First Lady's Conference on Women. At one point, Maria recited a Hopi Indian poem that had touched Lorrie deeply. It reads, in part:

There is a river flowing now very fast,
It is so great and swift that there are those who
    will be afraid,
They will try to hold onto the shore.
They will feel they are torn apart and will suffer
    greatly.
Know the river has its destination.
The elders say we must let go of the shore, push
    off into the middle of the river,
Keep our eyes open, and our heads above water.

Lorrie said this poem moved her to tears. She recognizes that all of us have to find the courage to leave the shore. That means leaving the crutch of our lifelong complaints and resentments, or our unhappiness over our upbringing or our bodies or whatever. It means no longer focusing negative energy on things beyond our control. It means looking beyond the safety of the familiar.

Lorrie loves the image of letting go of the shore, finding the middle of the river, and letting the river take us. It's a reminder that our lives are a combination of what we can control, what we can't, and the results of the choices we make.

The river analogy works in our marriage and it helps us cope with matters such as our financial difficulties.

"As long as we can keep our heads above the water," Lorrie says, "we can make it." It's a beautiful way of looking at life.

Lorrie and I don't always succeed in staying optimistic, but we have tried our best to live our lives in the middle of the river. Or else we're on our favorite hilltop, looking at the world below, reminding ourselves that anything is possible.

11

MANAGING THE SITUATION

Al Haynes.

Pilots mention his name with reverence.

On July 19, 1989, he was the captain of United Airlines Flight 232, a DC-10 traveling from Denver to Chicago. There were 296 passengers and crew on board.

When I was a facilitator of the crew resource management course (CRM), the story of that flight served as one of our most useful teaching tools. And personally, Flight 232 has taught me a great deal about flying—and about life.

After taking off from Denver, Flight 232 flew uneventfully for about eighty-five minutes. Then, soon after crossing into the airspace above Iowa, with the plane at thirty-seven thousand feet and the first officer, William Records, at the controls, an explosion was heard coming from the rear of the plane. The cause was soon apparent: The center engine had failed. Captain Haynes, who was approaching thirty thousand hours of flying experience, asked Dudley Dvorak, the second officer (flight engineer), to go through the engine failure checklist. As this was under way, the cockpit crew realized that all three hydraulic systems were losing pressure. Hydraulics are necessary to control this type of airplane. The first officer was having trouble controlling the aircraft.

Captain Haynes took the controls and saw he could turn the plane to the right but not the left. After the flight engineer announced to the passengers that an engine had failed, an off-duty United check pilot named Dennis Fitch, seated in the main cabin, came up front and offered to help. Captain Haynes welcomed him into the cockpit.

This type of emergency was so rare that there was no training for it, no checklist. It would later be determined that the odds of a simultaneous failure of three hydraulic systems approached a billion to one. But Captain Haynes played the hand he was dealt, and relied on his decades of experience to improvise and to lead. He and the others realized that the only way to control the airplane was to manipulate the throttles.

The four men in the cockpit flew like that for more than forty minutes, trying to brainstorm ways they might get the damaged airplane to the ground in one piece. In essence, they had forty minutes to learn a new way of flying an airplane.

Traditionally in the airline industry, there had been a steep hierarchy in cockpits, and first and second officers had been reluctant to offer many suggestions to a captain. The fact that Captain Haynes solicited and welcomed input that day helped the crew find ways to solve this unanticipated problem, and have a better chance of making it to a runway.

At first, air traffic controllers were going to send the crippled aircraft to Des Moines International Airport. But the plane was turning on its own, to the west, and so a decision was made to send it to Sioux City Gateway Airport. "I'm not going to kid you," Captain Haynes told the passengers. "It's going to be a very hard landing."

The cockpit voice recorder captured both the collaborative professionalism and the poignant camaraderie that eased their tension.

At one point, Dennis Fitch said, "I'll tell you what, we'll have a beer when this is all done."

Captain Haynes replied: "Well, I don't drink, but I'll sure as hell have one."

They approached the airport at a speed of 215 knots, descending at 1,600 feet per minute, as they tried to slow down by raising the nose. The pilots did a remarkable job of touching down near the beginning of the runway. It looked like they might make it.

Then the right wing struck the runway. Witnesses said the aircraft cartwheeled as it broke apart and into flames. There were 111 fatalities—some on impact, others from smoke inhalation—but 185 people survived that day because of the masterful work of Captain Haynes and his crew. (Though there were serious injuries, everyone in the cockpit lived.) An investigation later determined that a fatigue crack caused a fracture of the fan disk in the center engine.

In CRM training, Flight 232 is considered one of the best examples of a captain leading a team effort while being ultimately responsible for the decisions and the outcome. Captain Haynes turned to all the resources at his disposal on a plane in great jeopardy. Given what his crew was up against, this could well have been a crash with no survivors. Their work in the cockpit will be studied for generations.

I was honored to be contacted by Captain Haynes after my experience on Flight 1549. He has spent much of his life since the Sioux City accident speaking about it around the world. He has made more than

1,500 speeches, donating his fees or speaking pro bono. He talks about what the rest of us might learn from his experiences that day, focusing on the importance of communication, preparation, execution, cooperation, and the word he uses, "luck." He also talks about the sadness that he'll never shake regarding those on the plane who didn't make it.

He told me these speeches, which he dedicates to those who died on his flight, have been therapeutic for him. Speaking about safety issues has helped him cope with survivor's guilt. "My job was to get people from Point A to Point B safely," he said. "For a while afterward, I felt I didn't do my job."

Captain Haynes, now seventy-seven, was my age, fifty-eight, on the day of the Sioux City accident. He told me that beyond what his crew did, there were other favorable factors that saved lives: It was a clear day without much wind. The Iowa Air National Guard happened to be on duty there and rushed to help. Rescue crews had recently received training for handling the crash of a large jet. And just when his plane hit, both hospitals in town were in the middle of a shift change, meaning twice the medical personnel were available to treat the many injured survivors, including Captain Haynes. He was brought to the hospital with a head injury that required ninety-two stitches. He had a concussion and his left ear was almost cut off.

So many people involved that day stepped up aggressively to do what needed to be done. I always keep in mind a remark made by the fire chief at the Sioux City airport: "Either you manage the situation, or the situation will manage you."

In the years after the accident, Captain Haynes lost his oldest son in a motorcycle crash. His wife died of a rare infection. Then his daughter needed a bone marrow transplant. But, through all of this, he was buoyed to learn that his efforts on Flight 232 were not forgotten. When insurance wouldn't fully cover his daughter's procedure, hundreds of people, including survivors of the Sioux City crash, donated more than $500,000. His daughter even received donations from families who lost loved ones on Flight 232.

Captain Haynes told me he has continually seen the good in people, and they have helped him make peace with what he was able to do that day in 1989—and what he couldn't do. Understandably, he has wondered what would have happened if his crew could have kept the wings level and landed flat. But even had they been able to do that, the plane might have hit the runway and exploded.

When we talked a few weeks after Flight 1549, Captain Haynes told me to be prepared for some anxious thoughts. "I'm sure you'll feel there's something more you could have done," he said. "Everybody second-guesses themselves. We did, too, for a while. And then we decided there was nothing else we could have done." He had read a great deal about my flight, and told me he agreed with the decisions Jeff and I made in the cockpit. This meant a lot to me.

He also said that after Flight 1549, a few passengers from his flight got in touch with him, just to touch base and commiserate. Airline accidents are always reminders of past airline accidents. "It brought back memories for all of us," Captain Haynes told me.

He said he felt a kinship with me, given the traumas associated with both of our flights, and the ways in which we were tested. We talked of how we're members of a select group now. And then he gave me advice: "Wait until you're ready, and then go back to work. You're a pilot. You should be flying."

In CRM training, we also taught the details about United Airlines Flight 811, bound from Honolulu to Auckland, New Zealand, on February 24, 1989. It was a Boeing 747-122 with 337 passengers and a crew of eighteen.

At about 2:08 A.M., sixteen minutes after taking off from Honolulu, the forward cargo door blew out. The floor in the passenger cabin, above the door, caved in because of the change in pressure, and five rows of seats with nine passengers were sucked out of the jet and fell into the Pacific below. A huge hole was left in the cabin, and two of the engines were in flames, severely damaged by debris ejected from the plane during the incident.

The pilots, who had been climbing to just over twenty-two thousand feet, decided to make a 180-degree turn. Their hope was to make it back to Honolulu, seventy-two miles behind them. It would be a terrifying ride for passengers, as debris and baggage from damaged overhead bins swirled through the cabin. Some said it felt like a tornado.

Captain Dave Cronin, First Officer Al Slader, and Second Officer Randal Thomas knew that this emergency involved much more than just a loss of cabin pressurization. It also involved engine failures. With half their engines out, they had difficulty maintaining altitude that would be needed to make it back to Honolulu.

Slader used the fuel control switches to shut off the two engines, but opted not to pull the engine fire shutoff handles, which were designed to prevent further fires. He was procedurally required to pull those handles when engines are severely damaged, but he realized if he did so, two hydraulic pumps would be lost, which would affect the crew's ability to maintain control of the aircraft. So he did not pull them.

The pilots dumped fuel to make the plane lighter. The flight attendants had passengers put on life jackets and then told them to "Brace!" After landing, fire trucks put out the flames. Though 9 people had died in the wake of the cargo door explosion, 346 people survived the flight.

An investigation determined that the cause was a faulty switch or wiring in the cargo door control system, and problems with the design of the cargo door.

The crew acted heroically because they knew, from their deep knowledge of the systems on that plane, that they would have to improvise and modify procedures in order to deal with this unexpected emergency. They acted bravely in getting the plane safely to the ground.

As I studied that accident, I filed away the fact that I might one day have to rely on my systems knowledge, not only on a checklist. Not every situation can be foreseen or anticipated. There isn't a checklist for everything.

I've come across a number of people over the years who think that modern airplanes, with all their technology and automation, can almost fly themselves.

That's simply not true. Automation can lower the workload in some cases. But in other situations, using automation when it is not appropriate can increase one's workload. A pilot has to know how to use a level of automation that is appropriate.

I have long been an admirer of Earl Wiener, Ph.D., a former Air Force pilot who is now retired from the University of Miami's department of management science. He is renowned for his work in helping us understand aviation safety.

He once told me about an appearance he made at a forum in which another speaker's topic was "the role of the pilot in the automated cockpit." When it was Dr. Wiener's turn to speak, he noted, wryly but rightly, that the session should have been called "the role of automation in the piloted cockpit."

Whether you're flying by hand or using technology to help, you're ultimately flying the airplane with your mind by developing and maintaining an accurate real-time mental model of your reality—the airplane, the environment, and the situation. The question is: How many different levels of technology do you want to place between your brain and the control surfaces? The plane is never going somewhere on its own without you. It's always going where you tell it to go. A computer can only do what it is told to do. The choice is: Do I tell it to do something by pushing on the control stick with my hand, or do I tell it to do something by using some intervening technology?

The Airbus A320, the aircraft we were flying as Flight 1549, has a fly-by-wire system, which in essence means the flight controls are moved by sending electrical impulses, rather than having a direct mechanical link between the control stick in the cockpit and the control surfaces on the wings and tail. The fly-by-wire system keeps you from exceeding predetermined values, such as the degree of pitch (how low or high the plane's nose can be versus the horizon), the bank angle (how steep a turn you can make), and how fast or slow you can go.

Dr. Wiener worried, and I agree, that the paradox of automation is that it often lowers a pilot's workload when that load is already low. And it sometimes increases the workload in the cockpit when it is already high.

Take, for instance, a last-minute runway change. In the old days, you could easily tune your radio navigation receiver to the frequency for the approach to the different runway. Now it might take ten or twelve presses of buttons on the computer to arrange for a runway change.

For those who believe technology is the answer to everything, Dr. Wiener would offer data to prove that isn't the case. He said that automated airplanes with the highest technologies do not eliminate errors.

They change the nature of the errors that are made. For example, in terms of navigational errors, automation enables pilots to make huge navigation errors very precisely. Consider American Airlines Flight 965, a Boeing 757 flying from Miami to Cali, Colombia, on December 20, 1995. Because two different waypoints (defined points along a flight path) were given the same name and the flight management computer displayed the nearer one as the second choice of the two, the pilots mistakenly selected the more distant one, putting the plane on a collision course with a mountain. Just 4 of the 163 people on the plane survived.

Dr. Wiener is not antitechnology, and neither am I. But technology is no substitute for experience, skill, and judgment.

One thing that has always helped make the airline industry strong and safe is the concept that pilots call "captain's authority." What that means is we have a measure of autonomy—the ability to make an independent, professional judgment within the framework of professional standards.

The problem today is that pilots are viewed differently. Over the years, we've lost a good deal of respect from our management, our fellow employees, the general public. The whole concept of being a pilot has been diminished, and I worry that safety can be compromised as a result. People used to say that airline pilots were one step below astronauts. Now the joke is: We're one step above bus drivers, but bus drivers have better pensions.

Airline managers seem to second-guess us more often now. There are more challenges. Thirty years ago, it would be unheard of for a mechanic or ramp worker to vociferously disagree with a captain. Now it happens.

I know that some captains don't represent the best of us. There may be circumstances and times when it is appropriate to challenge a captain. But sometimes we are questioned because others in the airline system want the operation to go more smoothly or be more timely or less costly.

There was a scene in the 2002 movie *Catch Me If You Can* that made me think. Set in the 1960s, and based on a true story, the film stars Leonardo DiCaprio as a con man who at one point impersonates a Pan Am pilot. In this particular scene, DiCaprio's character is watching a handsome captain in full uniform walking into a hotel accompanied by several beautiful young Pan Am stewardesses. The front desk manager comes out from behind the counter to greet them, welcoming the captain and his crew back to the hotel. It's just a passing moment in the movie, but it perfectly encapsulates the high level of respect given to airline crews then. I almost had tears in my eyes watching that reminder of what the Golden Age of Aviation was like—and how much flight crews have lost since then.

A few years ago, in *Flying* magazine, I read a column written by an airline captain who was nearing retirement. He was remembering his earliest days as a pilot, and comparing those days with today, when all airline employees, including pilots, are judged on their ability to follow rules. "We were hired for our judgment," he wrote. "Now we are being evaluated on our compliance."

In many ways, it's good that all airlines are more standardized today. There are appropriate procedures and we are bound to follow them. These days there are virtually no cowboys in the skies, ignoring items on their checklists. At the same time, however, I am concerned that compliance alone is not sufficient. Judgment—like Al Slader's decision—is paramount.

The way the best pilots see it: A captain's highest duty and obligation is always to safety. As we say it: "We have the power of the parking brake." The plane will not move until we feel we can operate the aircraft safely.

With authority comes great responsibility. A captain needs leadership skills to take the individuals on his crew and make them feel and perform like a team. It's a heavy professional burden on the captain to know he may be called upon to tap into the depths of his experience, the breadth of his knowledge, and his ability to think quickly, weighing everything he knows while accounting for what he cannot know.

I long have had great respect for pilots such as Al Haynes, Al Slader, and many others. And I believe that my knowledge and understanding of their actions was of great help to me on Flight 1549 as I made decisions in those tense moments over New York City.

12

THE VIEW FROM ABOVE

No two airports are exactly alike. They're almost like fingerprints in that way. Each one has a different geometry, runway layout, and arrangement of taxiways and terminal buildings. Each one differs in its direction and distance from the city center, and proximity to other landmarks.

I've never counted how many different runways I've landed on. I couldn't tell you the exact number of cities I've seen from the air. But I try to pay attention to the specific details of a place, and to hold on to a mental picture of the view. It could be helpful the next time I return, even if it's years later.

When pilots fly regular routes to a certain city, we become very familiar with what the area's landmarks look like from the air. From as high as twenty-five or thirty thousand feet, we can identify the tallest buildings, the local stadiums, the nearest large bodies of water, the major highways. We know the configurations of the runways, the seasonal weather conditions, and, once on the ground, the best place to get a reasonably healthy lunch in the terminal.

Given the US Airways hub system, I've done a lot of flying into Charlotte, Pittsburgh, and Philadelphia, so takeoffs and landings in those cities are a pilot's equivalent of driving your car out of your driveway and through your neighborhood.

On so many flights, I find myself thinking the same thoughts: about how beautiful Earth is—both the natural and the man-made beauty—and how lucky we are to call it our home.

There are many parts of the country I enjoy flying over or into. Approaching St. Louis on a clear day, you can see the 630-foot-tall Gateway Arch from ten miles away and 30,000 feet up. If the sun is at the right angle, you'll find sunlight glistening off the edge of the arch.

Flying into Las Vegas, in the clear desert air, you can see the Strip from a good distance even in the daytime. At night, it's a line of some of the brightest lights on the continent, beckoning from eighty miles away.

Seattle is a gorgeous city to fly into. When I was a pilot at PSA, I would sometimes fly up to Seattle from Los Angeles, and I knew by memory the volcanoes in the Cascade Range heading north—Mount McLoughlin, Mount Bachelor, the Three Sisters, Mount Washington, Mount Jefferson, Mount Hood, Mount Adams, Mount Rainier. Each mountain would loom into view, one after the other.

I've flown over a lot of places in America—Montana, Idaho, the Dakotas—where you travel great distances without much evidence of human habitation. It's a lonely kind of beauty, but it can tug at you. I also like flying on the East Coast, where the population density is striking. There's a constant stream of lights between Washington, D.C., and Boston. From the air, it has almost become one continuous megalopolis.

Flying down to Ft. Lauderdale, I like passing over Cape Canaveral and seeing its three-mile-long runway. What a thrill it would be to land the shuttle there. Florida trips also have reminded me of how easily nature can tear apart hundreds of miles of human development. For years after a spate of hurricanes in 2004 and 2005, thousands of homes in South Florida had blue tarps covering their roofs. It was sobering to fly above that checkerboard carpet of blue squares, to see the destructive powers of wind and rain.

In the early 1990s, when I was lower on the seniority list, I had to pilot a lot of red-eye flights. On so many of those red-eyes, I got to see the northern lights again and again. Especially in the wintertime, there were nights when for the whole trip, west to east, the lights would fill the entire northern horizon. To me, these lights—formed by charged particles colliding in the earth's magnetosphere—looked like curtains billowing gently in the wind, with their folds swaying in and out. Sometimes, the lights were a deep magenta or cherry red. Other times, as the lights were cycling, they were lime green. Rather than looking like a curtain, these green lights sometimes looked like an old TV with the vertical hold not adjusted properly and the lines on the TV rolling from bottom to top. I felt privileged to be in a place, night after night, where I could see such scenes.

A few years ago, my schedule included regular trips to Bermuda, Jamaica, the Dominican Republic, Costa Rica, and Antigua, which were a lot more fun than landing in Charlotte for the 141st time. I loved approaching the islands during daylight. We'd come in over shallow turquoise water, with the white, sandy beaches and lush green mountains ahead of us.

I used to fly from Albany, New York, to LaGuardia, and we'd pass over West Point, a trip that would often jog memories for me. One winter, when I was a cadet at the Air Force Academy, I was sent to West Point for a week as part of an exchange program. On that visit, everything there felt gray to me: the stone walls of the old buildings, the winter sky, the cadets' uniforms. I ate in the cavernous cadet dining hall, where I was told that General Douglas MacArthur made his last visit to West Point. He had come back to his beloved alma mater in 1962 to give his famous "Duty, Honor, Country" speech. Flying over West Point on winter days decades later, I'd find myself thinking about that speech and wondering what the current cadets were doing at that particular moment.

My schedule takes me into and out of LaGuardia about fifteen times a year, and in my career, I've flown there hundreds of times. So I know the general landscape and landmarks of the area very well.

In the New York corridor, when the weather is good, controllers often tell us to fly toward a specific landmark on the ground. This use of "reporting points"—especially important when pilots are flying visually in addition to using instruments—is less common in some other areas of the country, where the landmarks aren't as large or well-known.

"Direct to the statue. Follow the river," controllers will tell us, which means fly toward the Statue of

Liberty and then follow the Hudson. Or they'll point us to the Verrazano-Narrows Bridge, at the mouth of upper New York Bay. "Direct to the Narrows."

If time permits, I'll allow myself to take a moment to appreciate the physical beauty of the New York landscape. Below me are millions of people in hundreds of thousands of structures. It's pretty dramatic.

On a cloudless day with good visibility, when I can clearly see "The Lady"—pilots' shorthand for the Statue of Liberty—I can often make out the flash of flame in her torch. Passing over the statue, I'm reminded of how I used to love reading an illustrated children's book to Kate and Kelly when they were young. The book was about the building of the statue, how the French people gave it to the United States as a gift, and about "The New Colossus," the Emma Lazarus poem engraved on a bronze plaque at the base. I enjoyed that children's book even more than the girls did, partly because I've always found that poem by Emma Lazarus to be so moving and evocative. I can recite much of it from memory: ". . . and her name Mother of Exiles. From her beacon-hand glows worldwide welcome; Her mild eyes command the air-bridged harbor . . . I lift my lamp beside the golden door!"

When the girls were little and I was on a trip, I'd mail them postcards so they could get a sense of where

I was. Sometimes, I'd also send postcards to their teachers to share with the class. I'd offer a few lines with my own observations about, say, the Liberty Bell in Philly or the famous statues of ducklings in Boston Public Garden. When I sent the girls postcards of the Statue of Liberty, I described the thrill I felt flying over it, and how I had thought of them and our shared bedtime book.

I wish I could bring Lorrie and the girls with me to see the country more often. One of the perks of working in the airline industry always has been our ability to have our families fly free or at a reduced fare. We can fly in coach without charge on US Airways if seats are available. On other airlines, we pay a percentage of the fare, usually between a quarter and half of the regular price.

In past eras, pilots easily took their spouses and kids on vacations and impulsive sightseeing jaunts. These days, however, with low fares ensuring that airplanes are almost always full, it's much harder to get seats. It's yet another result of airline deregulation. Our employee travel benefits are now of limited usefulness.

In 2001, for instance, I was able to get four seats on a flight to Orlando, so Lorrie and I were able to take the girls to Disney World. But then we had trouble getting seats on a flight home to San Francisco. We kept running back and forth to different terminals, schlepping all our luggage, trying to find a flight on any available airline.

Kate, then eight years old, eventually had enough. "Why don't we buy tickets like everyone else?" she asked. In her eyes, I wasn't a big-shot pilot impressing her with my perks. I was a cheap, harried father making her pull her suitcase all over the airport.

Mostly, we buy regular tickets for flights now, because the hassles and uncertainties of trying to use my employee travel benefits just aren't worth it.

I'd say our most memorable free trip as a family was to New York in December 2002, when the girls were nine and seven.

I had a four-day trip scheduled, and each night had a layover in Manhattan. Impulsively, I called Lorrie from Pittsburgh.

"Let's take the girls out of school," I told her. "I can get the three of you on the next red-eye to Pittsburgh, and from there we're going to take a little surprise vacation." It was an echo of the good old days, when my father would decide to pull my sister and me out of school for a trip to Dallas.

Lorrie and the girls agreed to come. They arrived early in the morning in Pittsburgh, and I was waiting for them at their gate. I piloted the next US Airways flight to LaGuardia, and I was able to get them seats on my plane.

I just loved having them on board. I did my usual welcome announcement, but with a twist. "Ladies and gentlemen, this is Captain Sullenberger, and Katie and Kelly, this is Dad. We are bound for New York's LaGuardia Airport . . ."

Lorrie later told me the girls giggled when I said that. They felt like everyone was smiling at them. It was a nice moment.

We got to New York and it was bitterly cold, but we had a terrific time. We took a ferry by the Statue of Liberty. It was just fifteen months after the attacks of September 11, and Liberty Island itself was still off-limits. That night we went to see *42nd Street* on Broadway.

The next day I piloted a flight from LaGuardia to New Orleans and back, and Lorrie and the girls stayed in New York. They went to Macy's and visited Santa Claus. They took a sightseeing bus tour of the city. They went to Ground Zero.

I made it back by nightfall, and we saw the Christmas tree at Rockefeller Center, and went ice-skating. Then we got tickets for the Rockettes' Christmas show at Radio City Music Hall. Kate and Kelly were wide-eyed at the splendor of the theater, and having taken dance lessons themselves, they loved how the dancers were arranged perfectly by height, and how they performed together as a chorus line with such precision.

The next day I had to pilot a flight from New York to Nassau. As I was leaving LaGuardia, a major snowstorm began. I got the plane deiced and flew down to the Bahamas, where it was eighty degrees. As usual, I was only able to step out into the sunshine briefly, when I walked down the stairs to the tarmac. After a quick turnaround, we flew back to New York that afternoon.

All the way back from Nassau, I checked the hourly weather reports for LaGuardia, and saw it was snowing in New York and visibility was down to a quarter of a mile. The forecast was that conditions would improve at our arrival time. But as we got closer, it looked like we might have to divert to Pittsburgh, our alternate airport.

When we arrived in the New York area, the visibility improved slightly, allowing us to land on a plowed but still-snow-covered runway.

As I was walking through the terminal, I stopped to look at the TV monitors showing the scheduled arrivals. In column after column, every flight from every city, A to Z, had the same notation: "canceled," "canceled," "canceled . . ." But when I got to the Ns, there was one flight, from Nassau, showing an on-time arrival. My flight.

Turned out, I was in the right place at the right time, and was able to arrive just as the weather improved. I got to the hotel when Lorrie and the kids were just about to go to dinner. I was struck by the sight of Kate and Kelly, standing in the lobby wearing beautiful wool winter overcoats with velvety collars. Kelly's was red. Katie's was green. They looked like pretty little dolls, dressed up for a walk in snowy Manhattan. I was grateful to have made it back to the city to see that vision of them, walking through the lobby and then into the night.

For the rest of our stay, Lorrie and I dragged the girls around—onto subways, into cabs. Everywhere we went, Kate and Kelly were two short suburban girls, lost in a sea of taller, city-savvy adults. By the end of the trip, Kate told us, "This has been a lot of fun, but I'm tired of all the hustle and bustle."

We were able to get four seats on a flight back to San Francisco, and this time, I sat back in coach with them, and we all looked out the window together, watching the continent go by.

For a pilot, LaGuardia is a more challenging environment than the average airport. The volume of traffic in the New York area makes it a complicated airspace, with so many planes vying for slots to take off or land. There are three major airports in close proximity—JFK, Newark, and LaGuardia—plus smaller facilities such as Westchester County Airport in White Plains and Teterboro in New Jersey. The radio frequencies are busier than those in many other places in the country. A great many voices are in your ears, and there's a lot going on around you that you need to be aware of.

Another issue is that at LaGuardia, the runways are short and surrounded by water. So you pretty much have to nail your landings, since there's not a lot of extra room if you don't. When landing, you want to put the airplane on the runway in the right place, because you'll need to have enough room to stop. You aim for the "touchdown zone," which begins a thousand feet beyond the start of the runway.

In the winter, of course, there are often weather conditions to be concerned about. And you have to be ready for flight delays as you wait your turn to get your plane deiced.

Still, despite all this, I enjoy flying out of LaGuardia. I like the challenge of it, and especially the view from the air—Central Park, the Empire State Building, the gorgeous homes and boats out on Long Island. I kind of enjoy the passengers out of LaGuardia, too. They often have a seasoned manner about them, and they're not always as tough as they seem.

It's true that a lot of passengers who board in New York can be very direct. They'll push the limits. But veteran flight attendants know that the way to deal with them is to be self-assured and push back a little. If passengers are firmly told the boundaries, they're generally OK with them.

When a passenger asks for two drinks at once, a flight attendant might smile and say, "Just a minute. I'll get to you. It's not all about you, you know! Didn't your mother teach you that?" If the flight attendant has the right, humorous delivery, a lot of passengers smile back and accept it. Flight attendants have told me: "When you want someone to turn off his computer for landing, you can ask him nicely, or you can say, 'OK now, that's enough of you and that laptop!' "

On midweek US Airways flights from "LGA," there are a lot of business travelers, and they can be savvy fliers. I often fly from LaGuardia to Charlotte, which has become a major banking center. So I might have a dozen or more bankers on each of those flights. There are always rows and rows of other frequent fliers, too, people who fly so often that they have a pretty good knowledge of the airline industry, the responsibilities of the crew, and the role passengers might have to play in an emergency.

In the case of Flight 1549, a Thursday-afternoon trip down to Charlotte, that would turn out to be fortuitous.

On January 15, 2009, the day of Flight 1549, the snow around LaGuardia had stopped earlier in the morning. It was cold and clear, with scattered clouds. Winds were out of the north, so we prepared to take off toward the north.

We were flying an Airbus A320–214, built in France by Airbus Industrie. The particular plane assigned to us, delivered to US Airways in 1999, had logged 16,298 flights before our takeoff. It had been airborne for 25,241 hours. The left engine had seen 19,182 hours of service, and the right engine had served for 26,466 hours. The most recent maintenance "A check" (which is done every 550 flight hours) had been forty days earlier. The plane had its annual C check (a comprehensive inspection) nine months earlier. These are common statistics for planes flown by commercial airlines in the United States.

With First Officer Jeff Skiles at the controls, we lifted off on the northeast runway, runway 4, about four seconds before 3:26 P.M. Along with the two of us in the cockpit, there were 150 passengers and our three flight attendants—Donna Dent, Doreen Welsh, and Sheila Dail.

As soon as we passed the end of the runway, the local controller at LaGuardia passed us off to the departure controller, Patrick Harten, who works at New York Terminal Radar Approach Control (TRACON) in Westbury, Long Island. Fourteen minutes earlier, he had been assigned to the LaGuardia departure radar position, which handles all departures from LaGuardia.

I radioed Patrick: "Cactus fifteen forty-nine, seven hundred, climbing five thousand." That meant we were passing through seven hundred feet, on the way to five thousand feet. Complying with our departure instructions, we had turned left to a heading of 360 degrees. On the magnetic compass that's due north.

Patrick responded: "Cactus fifteen forty-nine, New York departure radar contact. Climb and maintain one five thousand." He was telling us to climb to 15,000 feet.

I responded: "Maintain one five thousand, Cactus fifteen forty-nine."

As we climbed through 1,000 feet, Jeff commanded: "And flaps one, please." I repeated, "Flaps one," as I moved the flap lever from the 2 to the 1 detent while Jeff lowered the nose, shallowing our climb as we accelerated.

Next, Jeff said, "Flaps up, please, after takeoff checklist."

I responded, "Flaps up." I retracted the flaps, verified that all the items on the after takeoff checklist were done, and announced, "After takeoff checklist complete."

The takeoff portion of the flight was now complete, and we were transitioning to the climb portion of the flight by retracting the flaps. The flaps were needed for takeoff, but for our climb would only produce unnecessary drag. The airplane was in a clean configuration—with landing gear and flaps retracted—and we began our acceleration to 250 knots.

We continued climbing and accelerating. That incredible New York skyline was coming into view. Everything so far was completely routine.

13

SUDDEN, COMPLETE, SYMMETRICAL

We'd been in the air for about ninety-five seconds, and had not yet risen to three thousand feet when I saw them.

"Birds!" I said to Jeff.

The birds were ahead of us, in what probably was a V formation. Jeff had noticed them a fraction of a second before I uttered the word, but there was no time for either of us to react. Our airplane was traveling at 3.83 statute miles a minute. That's 316 feet per second. That means the birds were about a football field away when I first saw them. I barely blinked and they were upon us.

There were many large birds, a dozen or more, and I saw them in outline, with their wings extended straight out horizontally. We were flying so fast compared with the birds that it looked as if they weren't even moving.

I just saw, in an instant, the cylindrical dark outlines of their bodies. I'd later learn they were Canada geese, weighing anywhere from eight to eighteen pounds, with six-foot wingspans, and as is their way, they were flying within sight of one another at perhaps fifty miles an hour.

The cockpit windows on the Airbus A320 are large, and as I looked out the front, I saw the birds were everywhere, filling the windscreen. It was not unlike Alfred Hitchcock's *The Birds*. I thought later that I should have tried to duck in case the windshield cracked from the birds' impact, but there was no time.

The cockpit voice recorder captured my interchange with Jeff and the sounds in the cockpit:

> Sullenberger (3:27 and 10.4 seconds): *"Birds!"*
> Skiles (3:27:11): *"Whoa!"*
> (3:27.11.4): *Sound of thumps/thuds, followed by shuddering sound.*
> Skiles (3:27:12): *"Oh, shit!"*
> Sullenberger (3:27:13): *"Oh, yeah."*
> (3:27:13): *Sound similar to decrease in engine noise/frequency begins.*
> Skiles (3:27:14): *"Uh-oh."*

As the birds hit the plane, it felt like we were being pelted by heavy rain or hail. It sounded like the worst thunderstorm I'd ever heard back in Texas. The birds struck many places on the aircraft below the level of the windshield, including the nose, wings, and engines. The thuds came in rapid succession, almost simultaneously but a fraction of a fraction of a second apart.

I would later learn that Sheila and Donna, still strapped into their seats for takeoff, also felt the thuds.

"What was that?" Sheila asked.

"Might be a bird strike," Donna told her.

I had hit birds three or four times in my career and they had never even dented the plane. We'd make note of the strike in our maintenance logbook, make sure every piece of the airplane was unscathed, and that was it. I've long been aware of the risks, of course. About eighty-two thousand wildlife strikes—including deer, coyotes, alligators, and vultures—have been reported to the FAA since 1990. Researchers estimate that this is just a fifth of the actual number, since the great majority of strikes are never formally reported by pilots. Studies have shown that about 4 percent of strikes result in substantial damage to aircraft. In the past twenty years, wildlife strikes have resulted in 182 deaths and the destruction of 185 aircraft, according to the National Wildlife Research Center in Sandusky, Ohio.

At that moment on Flight 1549, a mere 2,900 feet above New York, I wasn't contemplating these statistics, however. What I focused on, extremely quickly, was that this situation was dire. This wasn't just a few small birds hitting the windshield or slapping hard against a wing and then falling to earth.

We were barely over 200 knots, that's 230 miles an hour, and immediately after the bird strike, I felt, heard, and smelled evidence that birds had entered the engines—both engines—and severely damaged them.

I heard the noise of the engines chewing themselves up inside, as the rapidly spinning, finely balanced machinery was being ruined, with broken blades coming loose. I felt abnormal, severe vibrations. The engines were protesting mightily. I'll never forget those awful, unnatural noises and vibrations. They sounded and felt BAD! And then I smelled a distinct odor—burning birds. The telltale air was being drawn from the engines into the cabin.

Within a few seconds, Jeff and I felt a sudden, complete, and bilaterally symmetrical loss of thrust. It was unlike anything I'd ever experienced in a cockpit before. It was shocking and startling. There's no other way to describe it. Without the normal engine noises, it became eerily quiet. Donna and Sheila would later tell me that in the cabin, it was as quiet as a library. The only remaining engine noise was a kind of rhythmic rumbling and rattling, like a stick being held against moving bicycle spokes. It was a strange windmilling sound from broken engines.

If you've got more than 40,000 pounds of thrust pushing your 150,000-pound plane uphill at a steep angle and the thrust suddenly goes away— completely—well, it gets your attention. I could feel the momentum stopping, and the airplane slowing. I sensed that both engines were winding down. If only one engine had been destroyed, the plane would be yawing, turning slightly to one side, because of the thrust in the still-working engine. That didn't happen. So I knew very quickly that this was an unparalleled crisis.

If we had lost one engine, we'd have maintained control of the airplane and followed the procedures for that situation. We'd have declared an emergency and told the controller about the loss of an engine, and received permission to land immediately at the most appropriate nearby airport. Then we would have told the flight attendants and passengers what was going on. It would be an emergency, but we would have almost certainly landed safely, probably at the airport in Newark, where the runways are longer than at LaGuardia.

The failure of even one engine had never happened to me before. Engines are so reliable these days that it is possible for a professional airline pilot to go an entire career without losing even one. I was headed for that perfect record before Flight 1549.

> Sullenberger (3:27:15): *"We got one roll—both of 'em rolling back."*
> (3:27:18): *Rumbling sound begins.*
> Sullenberger (3:27:18.5): *"Ignition, start."*
> Sullenberger (3:27:21.3): *"I'm starting the APU [auxiliary power unit]."*

Within eight seconds of the bird strike, realizing that we were without engines, I knew that this was the worst aviation challenge I'd ever faced. It was the most sickening, pit-of-your-stomach, falling-through-the-floor feeling I had ever experienced.

I knew immediately and intuitively that I needed to be at the controls and Jeff needed to handle the emergency checklist.

"My aircraft," I said to him at 3:27:23.2.

"Your aircraft," he responded.

This important protocol ensured that we both knew who was flying.

In the more common emergencies we train for, such as the loss of one engine, we would have time to go through our checklists and mull over solutions. In those cases, it is usually optimal for the first officer to fly so the captain can think about the situation, make decisions, and give direction.

Even in those early seconds, I knew this was an emergency that called for thinking beyond what's usually considered appropriate. As a rush of information came into my head, I had no doubts that it made the most sense for me to take the controls.

The reasons were clear to me. For one, I had greater experience flying the A320. Jeff was much newer to this type of plane. Also, all the landmarks I needed to see in order to judge where we might go were on my side of the airplane.

I also knew that since Jeff had just trained on the A320, he had more recent experience practicing the emergency procedures. He could more quickly find the right checklist out of about 150 checklists in our Quick Reference Handbook (QRH). He was the right man for that job.

After I took control of the plane, two thoughts went through my mind, both rooted in disbelief: *This can't be happening. This doesn't happen to me.*

I was able to force myself to set those thoughts aside almost instantly. Given the gravity of this situation, I knew that I had seconds to decide on a plan and minutes to execute it.

I was aware of my body. I could feel an adrenaline rush. I'm sure that my blood pressure and pulse spiked. But I also knew I had to concentrate on the tasks at hand and not let the sensations in my body distract me.

Jeff seemed to be equally on task. He was business-like, focused on what he had to do. He would later say his brain felt swelled "like when you have a bad head cold," but to me at the time, his voice and demeanor seemed unaffected. We both were very aware of how terrible this was. We just didn't waste time verbalizing this awareness to each other.

I've always kept in mind something said by astronaut John Young just before launch on a space mission. Asked if he was worried about the risks, or about the potential for catastrophe, he replied: "Anyone who sits on top of the largest hydrogen-oxygen fueled system in the world, knowing they're going to light the bottom, and doesn't get a little worried, does not fully understand the situation."

In our case, both Jeff and I clearly understood the gravity of our situation, and we were very concerned. Success would come if, at each juncture in the seconds ahead, we could solve the next problem thrown at us. Despite everything—the ruined plane, the sensations in my body, the speed with which we had to act—I had confidence that we could do it.

. . .

There are three general rules about any aircraft emergency. We learn them in our earliest lessons as pilots. And for those of us who served in the military, these rules are codified.

Maintain aircraft control.

Always make sure someone is flying the airplane, and is focused on maintaining the best flight path. No matter what else happens, you have to remember to fly the plane first, because if you don't, bad things can happen quickly.

There will be impulses to do other things: getting your mind around the particulars of the emergency, troubleshooting, finding the right checklists, talking to air traffic control. All of these things need to be done, but not at the expense of flying the airplane.

Analyze the situation and take proper action.

Through our training, we know the actions we should consider depend upon what systems have failed and how much time and fuel we have to deal with the situation. There are specific procedural steps, and we need to know them and be ready to take them.

Land as soon as conditions permit.

This means we have to factor in weather and runway conditions, the wind, the length and width of the runway, the emergency and rescue equipment available at the particular airport where a landing might be attempted, and all sorts of other factors. It is important to land quickly but with due consideration. How well will emergency crews at the closest airport be able to help? Does it make more sense to fly to another airport with better weather or facilities?

Those are the three basic rules. And there is a variation on these rules that pilots find easy to remember: "Aviate, navigate, communicate."

Aviate: Fly the plane. *Navigate:* Make sure your flight path is appropriate and that you're not flying off course. *Communicate:* Let those on the ground help you, and let those on the plane know what might be necessary to save their lives.

On Flight 1549, Jeff and I were doing all of these things almost simultaneously. We had no choice. That also meant we had to make sure that higher-priority tasks weren't suffering as we worked to accomplish the lower-priority tasks.

The first thing I did was lower the plane's nose to achieve the best glide speed. For all of us on board to survive, the plane had to become an efficient glider.

In the days that followed the Hudson landing, there was speculation in the media that all of my training as a glider pilot thirty-five years earlier had helped me on Flight 1549. I have to dispel that notion. The flight characteristics and speed and weight of an Airbus are completely different from the characteristics of the gliders I flew. It's a night-and-day difference. So my glider training was of little help. Instead, I think what helped me was that I had spent years flying jet airplanes and had paid close attention to energy management. On thousands of flights, I had tried to fly the optimum flight path. I think that helped me more than anything else on Flight 1549. I was going to try to use the energy of the Airbus, without either engine, to get us safely to the ground . . . or somewhere.

On Flight 1549, as we descended and I watched the earth came toward us faster than usual, the passengers did not immediately know how dire this was. They weren't flying the airplane, and they didn't have the training. Most probably, they couldn't put all these disparate cues into a worldview that would tell them the magnitude of our problem. The nature of the emergency and the extreme time compression forced Jeff and me to focus our attention on the highest-priority tasks, so there was no time to make any verbal contact with those in the cabin, even the flight attendants.

In the cockpit, Jeff and I never made eye contact, but from the few words he spoke and his overall demeanor and body language, I had the clear sense that he was not panicked. He was not distracted. He was working quickly and efficiently.

Sullenberger (3:27:28): *"Get the QRH . . . Loss of thrust on both engines."*

Jeff grabbed the Quick Reference Handbook to find the most appropriate procedure for our emergency. The QRH book is more than an inch thick, and in previous editions, it had helpful numbered tabs sticking out of the edge of it. That made it easier for us to find the exact page we needed. You could hold it in your left hand and use it like an address book, grazing over the numbered tabs with your right hand before turning to the tab for, say, Procedure number 27.

In recent years, however, in a cost-cutting move, US Airways had begun printing these booklets without the numbered tabs on the edge of the pages. Instead, the number of each procedure was printed on the page itself, requiring pilots to open the pages and thumb through them to get to the right page.

On Flight 1549, as Jeff turned quickly through the pages of his QRH without tabs, it likely took him a few extra seconds to find the page he needed with the proper procedure. I told this to the National Transportation Safety Board in my testimony given in the days after the accident.

We were over the Bronx at that point and I could see northern Manhattan out the window. The highest we ever got was just over three thousand feet, and now, still heading northwest, we were descending at a rate of over one thousand feet per minute. That would be equivalent to an elevator descending two stories per second.

Twenty-one and a half seconds had passed since the bird strike. I needed to tell the controller about our situation. I needed to find a place to put the plane down quickly, whether back at LaGuardia or somewhere else. I began a left turn, looking for such a place.

"Mayday! Mayday! Mayday! . . ."

That was my message—the emergency distress signal—to Patrick Harten, the controller, just after 3:27:32.9. My delivery was businesslike, but with a sense of urgency.

Patrick never heard those words, however, because while I was talking, he was making a transmission of his own—to me. Once someone keys his microphone, he can't hear what's being said to him on the same frequency. While Patrick was giving me a routine direction—"Cactus fifteen forty-nine, turn left heading two seven zero"—my "Mayday" message was going no farther than our cockpit.

I didn't know that Patrick hadn't heard me and that I hadn't heard him. This is a regular and problematic issue in communications between controllers and pilots. When two people transmit simultaneously, they not only block each other, but they also sometimes prohibit others nearby from hearing certain transmissions. "Anti-blocking" devices have been invented that allow aircraft radios to detect when someone else is transmitting. That way, once a radio senses another transmission, it can prevent your radio from transmitting so you don't block someone else. We could certainly use such devices or similar technology in our cockpits. All pilots have stories. There have been times when a pilot will bump his radio's button, and for a few minutes, those of us in planes on the same frequency hear only background noise from that pilot's cockpit. We can't hear the controller. It is a potentially hazardous situation that has not been resolved because airlines and other operators have chosen not to adopt anti-blocking technology, and the FAA has not mandated it.

Patrick's transmission lasted about four seconds, and when he released his transmit button, he heard the rest of my transmission: ". . . This is, uh, Cactus fifteen thirty-nine. Hit birds. We've lost thrust in both engines. We're turning back towards LaGuardia."

I had gotten the flight number wrong. Later, when I heard the tape, I detected a higher stress level in my voice. My voice quality was slightly raspy, slightly higher pitched. No one else might have noticed, but I could hear it.

Patrick, a thirty-four-year-old controller, had worked many thousands of flights in his ten years on the job, and had a reputation for being careful and diligent.

He had assisted a few jets with failures of one engine, though none to the point where the plane had become a glider. He worked to get these flights back to the ground as quickly as possible, and in each case, the planes landed without incident. Like other controllers, he took pride in the fact that he had never failed in his attempts to help a plane in distress get safely to a runway.

In Patrick's previous emergencies, he had remained calm and acted intelligently.

Once, he had a plane coming in from overseas. There was bad weather that day, and the plane had been held in holding patterns. Eventually, it had enough fuel to last just thirty more minutes. The plane was almost twenty minutes from the airport. If a new weather problem developed, or there was a further traffic delay, the plane could run out of fuel. Knowing there was no margin for error, Patrick had to pull another aircraft from its final approach, and slot in the plane with low fuel. He oversaw the rearranging of a jigsaw puzzle in the sky, and was able to help the plane land without incident.

About fifteen times in his career, Patrick had pilots tell him that their planes had just hit birds. The worst bird strike he had ever handled before Flight 1549 involved a cracked windshield. Patrick had helped that airplane return to LaGuardia safely.

Patrick certainly had his share of experiences with emergencies. But like almost every controller working in the world today, he had never been in a situation where he was guiding a plane that had zero thrust capability.

In the case of Flight 1549, Patrick knew he had to act quickly and decisively. He made an immediate decision to offer us LaGuardia's runway 13, which was the closest to our current position. At that moment, we were still heading away from LaGuardia and descending rapidly.

He made no comment, of course, about the seriousness of the condition of our plane. He just responded.

"OK, uh," he radioed back to me. "You need to return to LaGuardia. Turn left heading of, uh, two two zero."

"Two two zero," I acknowledged, because I knew all my options lay to my left. In the left turn, I would have to choose one, and the option I chose would determine the ultimate heading I would fly.

From the cockpit voice recorder:

Skiles (3:27:50): *"If fuel remaining, engine mode selector, ignition. Ignition."*

Sullenberger (3:27:54): *"Ignition."*

Skiles (3:27:55): *"Thrust levers, confirm idle."*

Sullenberger (3:27:58): *"Idle."*

Skiles (3:28:02): *"Airspeed optimum relight. Three hundred knots. We don't have that."*

Flight warning computer (3:28:03): *Sound of single chime.*

Sullenberger (3:28:05): *"We don't . . ."*

Patrick immediately contacted the tower at LaGuardia telling them to clear all runways. "Tower, stop your departures. We got an emergency returning."

"Who is it?" the tower controller asked.

"It's fifteen twenty-nine," said Patrick, also getting the flight number wrong in the stress of the moment. "Bird strike. He lost all engines. He lost the thrust in the engines. He is returning immediately."

Losing thrust in both engines is so rare that the LaGuardia controller didn't fully recognize what

Patrick had just told him. "Cactus fifteen twenty-nine. *Which* engine?" he asked.

Patrick replied: "He lost thrust in *both* engines, he said."

"Got it," said the LaGuardia controller.

You won't hear it on the tape, because none of the controllers said it out loud, but in their minds they thought they were working a flight that would likely end very tragically.

Worldwide, airliners lose thrust in all engines so rarely that a decade can pass between occurrences. Usually, planes lose thrust in all engines only when they fly through a volcanic ash cloud or there is a fuel problem. And in the case of a volcanic ash encounter, the pilots have had enough time to get their engines to restart once clear of the ash cloud. Because they were at a high enough altitude—well above thirty thousand feet—there was time to go through their procedures and work on a solution, to get at least one engine going again.

In the case of Flight 1549, however, even if we were as high as the moon, we would never have gotten our engines restarted because they were irreparably damaged. Given the vibrations we felt coming from the engines, and the immediate loss of thrust, I was almost certain we'd never get the engines working again. And yet I knew we had to try.

So while Jeff worked diligently to restart at least one engine, I focused on finding a solution. I knew we had fewer than a handful of minutes before our flight path would intersect the surface of the earth.

I had a conceptual realization that unlike every other flight I'd piloted for forty-two years, this one probably wouldn't end on a runway with the airplane undamaged.

14

GRAVITY

Less than a minute had passed since the bird strikes ruined the engines on Flight 1549. At his radar position out on Long Island, Patrick, the controller, was still hoping he could get us to a runway at LaGuardia.

Controllers guide pilots to runways. That's what they do. That's what they know best. So he wasn't going to abandon that effort until every option had been exhausted. He figured that even in this most dire of situations, most pilots would have tried to make it back to LaGuardia. He assumed that would be my decision, too.

Five seconds after 3:28 P.M., which was just 32 seconds after I first made Patrick aware of the emergency, he asked me: "Cactus fifteen twenty-nine, if we can get

it to you, do you want to try to land runway one three?"
Patrick was offering us the runway at LaGuardia that
could be reached by the shortest path.

"We're unable," I responded. "We may end up in
the Hudson."

I knew intuitively and quickly that the Hudson River
might be our only option, and so I articulated it. It felt
almost unnatural to say those words, but I said them.
In his seat to my right, Jeff heard me and didn't comment. He was busy trying to restart the engines. But
he later told me he silently acknowledged my words in
his own head, thinking I might have been right. The
Hudson could turn out to be our only hope.

We both knew that our predicament left us few
choices. We were at a low altitude, traveling at a low
speed, in a 150,000-pound aircraft with no engines.
Put simply: We were too low, too slow, too far away,
and pointed in the wrong direction, away from the
nearby airports.

If there had been a major interstate highway without overpasses, road signs, or heavy traffic, I could
have considered landing on it. But there are very few
stretches of interstate in America without those barri-
ers these days, and certainly none of them are in New
York, the nation's largest metropolis. And, of course,
I didn't have the option of finding a farmer's field that

might be long enough and level enough. Not in the Bronx. Not in Queens. Not in Manhattan.

But was I really ready to completely rule out LaGuardia?

Looking out the window, I saw how rapidly we were descending. My decision would need to come in an instant: Did we have enough altitude and speed to make the turn back toward the airport and then reach it before hitting the ground? There wasn't time to do the math, so it's not as if I was making altitude-descent calculations in my head. But I was judging what I saw out the window and creating, very quickly, a three-dimensional mental model of where we were. It was a conceptual and visual process, and I was doing this while I was flying the airplane as well as responding to Jeff and Patrick.

I also thought quickly about the obstacles between us and LaGuardia—the buildings, the neighborhoods, the hundreds of thousands of people below us. I can't say I thought about all of this in any detail. I was quickly running through a host of facts and observations that I had filed away over the years, giving me a broad sense of how to make this decision, the most important one of my life.

I knew that if I chose to turn back across this densely populated area, I had to be certain we could make it. Once I turned toward LaGuardia, it would be an irrevocable choice. It would rule out every other option. And attempting to reach a runway that was unreachable could have had catastrophic consequences for everyone on the airplane and who knows how many people on the ground. Even if we made it to LaGuardia and missed the runway by a few feet, the result would be disastrous. The plane would likely tear open and be engulfed in flames.

I also considered the fact that, no matter what, we'd likely need a serious rescue effort. I knew that the water rescue resources at LaGuardia were a tiny fraction of those available on the Hudson between Manhattan and New Jersey. It would take much longer for rescue workers at LaGuardia to reach us and then help us if we tried for the runway and missed it.

And even if we could remain airborne until we were over a runway, there were potential hazards. Jeff would have had to stop trying to restart the engines, and instead turn his attention to preparing for a landing on a runway. I'd have to be able to appropriately manage our speed and altitude to try to touch down safely.

We had hydraulic power to move the flight control surfaces, but we didn't know if we'd be able to lower the landing gear and lock it into position. We might have had to use an alternate method—one in which gravity lowers the landing gear—and that would require another checklist for Jeff to attend to.

We would have had to be able to align the aircraft's flight path exactly with a relatively short runway, touch down at an acceptable sink rate, and maintain directional control throughout the landing to make sure we didn't run off the runway. Then we'd have to make sure the brakes would work, stopping before the end of the runway. Even then, would the airplane remain intact? There could be fire, smoke inhalation, and trauma injuries.

I also knew that if we turned toward LaGuardia and were unable to reach the airport, there would be no open stretch of water below us until Flushing Bay. And even if we were forced to ditch in that bay, near LaGuardia, and land in one piece, I feared that many on the aircraft would perish afterward. Rescuers there have access to just a few outboard motorboats, and it probably would have taken them too long to get to the aircraft, and too many trips to carry survivors to shore.

The Hudson, even with all the inherent risks, seemed more welcoming. It was long enough, wide enough, and, on that day, it was smooth enough to land a jet airliner and have it remain intact. And I knew I could fly that far.

I was familiar with the *Intrepid*, the famed World War II aircraft carrier that is now the Intrepid Sea-Air-Space Museum. It is docked on the Hudson by North River Pier 86, at Forty-sixth Street on the West Side of Manhattan. On my visit to the museum a few years earlier, I had noticed there were a lot of maritime resources nearby. I'd seen all the boat traffic there. So it did occur to me that if we could make it safely into the Hudson near the *Intrepid*, there would be ferries and other rescue boats close by, not to mention large contingents of the city's police and ambulance fleets just blocks away.

Patrick, the controller, was less optimistic about ditching in the Hudson. He assumed no one on the plane would survive it. After all, flight simulators that pilots practice on don't even have an option to land in water. The only place we train on ditching is in the classroom.

Before Patrick could even get back to me, he had another plane he had to attend to. "Jetlink twenty-seven sixty," he said, "turn left, zero seven zero." Then back to me, still trying to steer me to LaGuardia, he said: "All right, Cactus fifteen forty-nine, it's gonna be left traffic for runway three one."

I was firm. "Unable."

From everything I saw, knew, and felt, my decision had been made: LaGuardia was out. Wishing or hoping otherwise wasn't going to help.

Inside the cockpit, I heard the synthetic voice of the Traffic Collision Avoidance System issuing an aural warning: "Traffic. Traffic."

Patrick asked: "OK, what do you need to land?"

I was looking out the window, still going through our options. I didn't answer, so Patrick again offered LaGuardia. "Cactus fifteen twenty-nine, runway four's available if you wanna make left traffic to runway four."

"I'm not sure we can make any runway," I said. "Uh, what's over to our right? Anything in New Jersey? Maybe Teterboro?"

Teterboro Airport in New Jersey's Bergen County is called a "reliever airport," and handles a lot of the New York area's corporate and private jet traffic. Located twelve miles from midtown Manhattan, it has more than five hundred aircraft operations a day.

"You wanna try and go to Teterboro?" Patrick asked.

"Yes," I said. It was 3:29 and three seconds, still less than a minute after I had first made Patrick aware of our situation.

Patrick went right to work. His radar scope had a touch screen, giving him the ability to call any one of about forty different vital landlines. With one movement of his finger, he was able to get through to the air traffic control tower at Teterboro. "LaGuardia departure," he said, introducing himself, "got an emergency inbound." Later, listening to the recording of the conversation, Patrick could hear the distress in his voice. But he remained direct and professional.

The controller at Teterboro responded: "Okay, go ahead."

Patrick could see on his radar screen that I was about nine hundred feet above the George Washington Bridge. "Cactus fifteen twenty-nine over the George Washington Bridge wants to go to the airport right now," he said.

Teterboro: "He wants to go to our airport. Check. Does he need any assistance?" The Teterboro controller was asking if fire trucks and emergency responders should leave their stations immediately.

Patrick answered: "Ah, yes, he, ah, he was a bird strike. Can I get him in for runway one?"

Teterboro: "Runway one, that's good."

They were arranging for us to land on the arrival runway, because it was the easiest to clear quickly of traffic.

Patrick was doing several smart and helpful things in dealing with our flight, which, in retrospect, leaves me very grateful. For starters, he didn't make things more complicated and difficult for Jeff and me by over-loading us.

In emergencies, controllers are supposed to ask pilots basic questions: "How much fuel do you have remaining?" "What is the number of 'souls on board'?" That would be a count of passengers and crew so rescue workers could know how many people to account for.

"I didn't want to pester you," Patrick later told me. "I didn't want to keep asking, 'What's going on?' I knew I had to let you fly the plane."

Also, in order to save seconds and not have to repeat himself, he left the phone lines open when he called the controllers at other airports, so they could hear what he was saying to me and what I was saying to him. That way he wouldn't have to repeat himself. The improvising he did was ingenious.

Patrick's conscious effort not to disturb me allowed me to remain on task. He saw how quickly we were descending. He knew I didn't have time to get him passenger information or to answer any questions that weren't absolutely crucial.

The transcripts of our conversation also show how Patrick's choice of phrasing was helpful to me. Rather than telling me what airport I had to aim for, he asked me what airport I wanted. His words let me know that he understood that these hard choices were mine to make, and it wasn't going to help if he tried to dictate a plan to me.

Through all my years as a commercial pilot, I had never forgotten the aircrew ejection study I had learned about in my military days. Why did pilots wait too long before ejecting from planes that were about to crash? Why did they spend extra seconds trying to fix the unfixable? The answer is that many doomed pilots feared retribution if they lost multimillion-dollar jets. And so they remained determined to try to save the airplane, often with disastrous results.

I had never shaken my memories of fellow Air Force pilots who didn't survive such attempts. And having the details of that knowledge in the recesses of my brain was helpful in making those quick decisions on Flight 1549. As soon as the birds struck, I could have attempted a return to LaGuardia so as not to ruin a US Airways aircraft by attempting a landing elsewhere. I could have worried that my decision to ditch the plane would be questioned by superiors or investigators. But I chose not to.

I was able to make a mental shift in priorities. I had read enough about safety and cognitive theory. I knew about the concept of "goal sacrificing." When it's no longer possible to complete all of your goals, you sacrifice lower-priority goals. You do this in order to perform and fulfill higher goals. In this case, by attempting a water landing, I would sacrifice the "airplane goal" (trying not to destroy an aircraft valued at $60 million) for the goal of saving lives.

I knew instinctively and intuitively that goal sacrificing was paramount if we were to preserve life on Flight 1549.

It took twenty-two seconds from the time I considered and suggested Teterboro to the time I rejected the airport as unreachable. I could see the area around Teterboro moving up in the windscreen, a sure sign that our flight path would not extend that far.

"Cactus fifteen twenty-nine, turn right two eight zero," Patrick told me at 3:29 and twenty-one seconds. "You can land runway one at Teterboro."

"We can't do it," I answered.

"OK, which runway would you like at Teterboro?" he asked.

"We're gonna be in the Hudson," I said.

Patrick had heard me just fine. But he asked me to repeat myself.

"I'm sorry, say again, Cactus," he said.

"I simply could not wrap my mind around those words," Patrick would later explain in testimony before Congress. "People don't survive landings on the

Hudson River. I thought it was his own death sentence. I believed at that moment I was going to be the last person to talk to anyone on that plane."

As he spoke to me, Patrick couldn't help but think about Ethiopian Airlines Flight 961, hijacked in 1996. A Boeing 767-260ER, it ran out of fuel and attempted to land in the Indian Ocean just off the coast of the island nation of Comoros. The aircraft's wingtip struck the water first, and it spun violently and broke apart. Of the 175 people on board, 125 died from either the impact or drowning. Photos and video of the cartwheeling Boeing 767-260ER are easy to find on the Internet. "That's the picture I had in my head," Patrick said.

Patrick continued to talk to me, but I was too busy to answer. I knew that he had offered me all the assistance that he could, but at that point, I had to focus on the task at hand. I wouldn't be answering him.

As we descended toward the Hudson, falling below the tops of New York's skyscrapers, we dropped off his radar. The skyline was now blocking transmissions.

Patrick tried desperately to find a solution that would keep us out of the water. At 3:29:51: "Cactus, uh, Cactus fifteen forty-nine, radar contact is lost. You also got Newark airport off your two o'clock in about seven miles."

At 3:30:14: "Cactus fifteen twenty-nine, uh, you still on?"

He feared we had already crashed, but then we flickered back onto his radar scope. We were at a very low altitude, but because we had returned to radar coverage, he hoped against hope that maybe we had regained use of one of our engines.

At 3:30:22 he said: "Cactus fifteen twenty-nine, if you can, uh, you got, uh, runway two nine available at Newark. It'll be two o'clock and seven miles."

There was no way to answer him. By then, we were just 21.7 seconds from landing in the river.

Had we lost one engine instead of two, Jeff and I would have had more time to analyze things and to communicate with the crew and passengers. We could have had the flight attendants prepare the cabin. We could have asked air traffic control to help us determine the best plan for our return. But on Flight 1549, there was much we couldn't do because everything was so terribly time-compressed.

Many of the passengers had felt the bird strike. They heard the sound of the birds thumping against the plane, and the disturbing bangs that preceded the failing of the engines. They saw some smoke in the cabin, and like me, they could smell the incinerated birds.

Actually, more accurately, the birds were liquefied into what is referred to as "bird slurry."

I have heard the stories of what the passengers were going through while I was so occupied in the cockpit. Many would later write notes to me, sharing their personal recollections. Others gave media interviews that I found moving and haunting.

There was former U.S. Army Captain Andrew Gray, who had completed two tours of duty in Afghanistan, and was traveling on Flight 1549 with his fiancée, Stephanie King. As the plane descended, Andrew and Stephanie kissed and told each other "I love you." As they described it, they "accepted death together."

John Howell, a management consultant from Charlotte, thought about how he was his mother's only surviving son. His brother, a firefighter, died at the World Trade Center on September 11, 2001. John later told reporters that as Flight 1549 descended, "The only thing I was thinking was, 'If I go down, my mother's not going to survive this.'"

In 12F, a window seat just behind the wing emergency exit, forty-five-year-old Eric Stevenson was experiencing an awful feeling of déjà vu. On June 30, 1987, he had been on Delta Air Lines Flight 810, a Boeing 767, traveling from Los Angeles to Cincinnati. Shortly after takeoff, as the plane was climbing over the Pacific before turning east, one of the pilots had mistakenly shut down both engines. He had done this inadvertently because of the way the engine control panel was designed and the proximity of similar engine control switches. The plane began descending from 1,700 feet, while passengers quickly donned life jackets and expected the worst. Hearing some passengers crying around him, Eric decided to take out one of his business cards and write the words "I love you" to his parents and sister. He shoved it in his pocket, figuring he was likely to die and the note might be found on his body. Then just 500 feet above the water, the passengers felt a massive burst of thrust and the plane jolted forward with full force. The pilots had restarted the engines. The flight continued to Cincinnati, its cabin littered with life preservers. After that incident, Boeing redesigned the engine control panel to prevent a recurrence.

That near-death experience led Eric to take a year off from work so he could travel the world, and every year after that, he found ways to solemnly mark the anniversary of the incident. He said it planted the seeds for his eventual move to Paris, where he continues to work as a marketing manager for Hewlett-Packard. It was while visiting the United States in January 2009 that he ended up as a passenger on Flight 1549. Sitting in 12F, looking out the window, he couldn't believe he was on another airplane without working engines.

And so he again took out a business card and wrote "Mom and Jane, I love you." He shoved it into his right front pocket and thought to himself, "This will probably get separated from my body if the cabin disintegrates." But he felt a measure of comfort knowing he had taken this step. "It was the maximum I could do," he later told me. "All of us were completely at the mercy of the two of you in the cockpit. It was a helpless feeling, knowing there was nothing we could do about the situation. So I did the only thing I could do. With the plane going down, I wanted my family to know I was thinking about them at the very last moment."

As the plane descended, Eric didn't feel panic, but he did feel the same sadness he experienced at age twenty-three, in that Boeing 767 over the Pacific. On our flight, he recalled, he had the same clear thought: "This could be the end of my life. In ten or twenty seconds, I will be on the other side, whatever the other side will be."

The cabin was very quiet. A few people made phone calls or sent text messages to their loved ones. I'm told some were saying their prayers. Others would say they were making peace with the situation. If they were going to die, they said, there was nothing they could do about it, and so they tried to accept it.

Some of the passengers later told me that they were glad I didn't give them too many details. That would have made them even more frightened.

It wasn't until about ninety seconds before we hit the water that I spoke to the passengers.

I wanted to be very direct. I didn't want to sound agitated or alarmed. I wanted to sound professional.

"This is the captain. Brace for impact!"

I knew I had to make an announcement to the passengers to brace. We're taught to use that word. "Brace!" Saying it not only can help protect passengers from injury on touchdown, but it is also a signal to the flight attendants to begin shouting their commands. Even in the intensity of the moment, I knew I had to choose my words very carefully. There was no time to give the flight attendants a more complete picture of the situation we faced. So my first priority was to prevent passenger injury on impact. I did not yet know how well I'd be able to cushion the touchdown. I said "brace" and then chose the word "impact" because I wanted passengers to be prepared for what might be a hard landing.

The flight attendants—Sheila, Donna, and Doreen—immediately fell back on their training. All the cockpit doors have been hardened since the September 11 attacks, so it's more difficult to hear what's going on in the cabin. Still, through that thicker door, I could hear Donna and Sheila, who were up front, shouting their commands in response to my announcement, almost in unison, again and again: "Brace, brace! Heads down! Stay down! Brace, brace! Heads down! Stay down!"

As I guided the plane toward the river, hearing their words comforted and encouraged me. Knowing that the flight attendants were doing exactly what they were supposed to do meant that we were on the same page. I knew then that if I could deliver the aircraft to the surface intact, Donna, Doreen, and Sheila would get the passengers out the exit doors and the rescue could begin. Their direction and professionalism would be keys to our survival, and I had faith in them.

From the cockpit voice recorder:

Sullenberger (3:29:45): *"OK, let's go put the flaps out, put the flaps out..."*
Enhanced Ground Proximity Warning System synthetic voice (3:29:55): *"Pull up. Pull up. Pull up. Pull up. Pull up. Pull up."*
Skiles (3:30:01): *"Got flaps out!"*
Skiles (3:30:03): *"Two hundred fifty feet in the air."*

The plane continued to descend, and it was as if the bluffs along the Hudson and the skyscrapers on both sides of the shoreline had come up to meet us. As Jeff would later describe it: "It felt as if we were sinking into a bathtub." The river below us looked cold.

Ground Proximity Warning System synthetic voice (3:30:04): *"Too low. Terrain."*

Ground Proximity Warning System (3:30:06): *"Too low. Gear."*

Skiles (3:30:06): *"Hundred and seventy knots."*

Skiles (3:30:09): *"Got no power on either one. Try the other one."*

Radio from another plane (3:30:09): *"Two one zero, uh, forty-seven eighteen. I think he said he's goin' in the Hudson."*

Enhanced Ground Proximity Warning System synthetic voice (3:30:15): *"Caution, terrain!"*

Skiles (3:30:16): *"Hundred and fifty knots."*

Skiles (3:30:17): *"Got flaps two, you want more?"*

Sullenberger (3:30:19): *"No, let's stay at two."*

Sullenberger (3:30:21): *"Got any ideas?"*

Skiles (3:30:23): *"Actually, not."*

Enhanced Ground Proximity Warning System synthetic voice (3:30:23): *"Caution, terrain."*

Enhanced Ground Proximity Warning System synthetic voice (3:30:24): *"Terrain, terrain. Pull up.*

Pull up." ["Pull up" repeats until the end of the recording.]
Sullenberger (3:30:38): *"We're gonna brace!"*

I did not think I was going to die. Based on my experience, I was confident that I could make an emergency water landing that was survivable. That confidence was stronger than any fear.

Lorrie, Kate, and Kelly did not come into my head, either. I think that was for the best. It was vital that I be focused, and that I allow myself no distractions. My consciousness existed solely to control the flight path.

As we came in for a landing, without thrust, the only control I had over our vertical path was pitch—raising or lowering the nose of the plane. My goal was to maintain a pitch attitude that would give the proper glide speed. In essence, I was using the earth's gravity to provide the forward motion of the aircraft, slicing the wings through the air to create lift.

My flight instruments were still powered. I could see the airspeed indication. If I was slower than I needed to be, I slightly lowered the nose. If I felt we were going too fast, I raised the nose.

As a fly-by-wire airplane, the Airbus has some flight envelope protections, which means the flight control computers interpret the pilot's sidestick inputs. Unlike more conventional aircraft, the Airbus does not provide the pilot with natural cues or "feel" that speed is changing, which would normally help the pilot maintain constant speed. But one of the fly-by-wire protections when flying at low speeds is that regardless of how hard the pilot pulls back on the sidestick, the flight control computers will not allow him to stall the wings and lose lift.

Compared with a normal landing, our rate of descent was much greater, since we had no engine thrust. Our landing gear was up, and I tried to keep the wings level to avoid cartwheeling when we hit the water. I kept the nose up.

My focus had narrowed as we descended, and now I was looking in only two places: the view of the river directly ahead and, inside the cockpit, the airspeed display on my instruments. Outside-inside-outside-inside.

It was only about three minutes since the bird strike, and the earth and the river were rushing toward us. I was judging the descent rate and our altitude visually. At that instant, I judged it was the right time. I began the flare for landing. I pulled the sidestick back, farther back, finally full aft, and held it there as we touched the water.

We landed and slid along the surface in a slightly nose-up attitude. The rear of the plane hit much harder than the front. Those in the back felt a violent impact. Those in front felt it as more of a hard landing.

We slowed down, leveled out, and then came to a stop as the river water splashed over the cockpit windows. I would later learn that I had achieved most of the parameters I attempted: The plane had landed with the nose at 9.8 degrees above the horizon, the wings were exactly level, and we were flying at 125.2 knots, just above the minimum speed for that configuration. The rate of descent, however, even with full aft stick commanding full nose up, could not be arrested as much as I would have liked.

Within a second or two, we returned to the slightly nose-up attitude and the plane was floating. The skyline of New York presented itself from sea level.

Jeff and I turned to each other and, almost in unison, said the same thing.

"That wasn't as bad as I thought."

Still, we knew that the hardest part of this emergency might still be ahead. There were 155 passengers and crew members on a plane that might soon be sinking.

15

ONE HUNDRED FIFTY-FIVE

The water landing certainly wasn't as bad as Jeff and I knew it could have been. We didn't cartwheel when we touched down. The aircraft remained intact. The fuel didn't ignite. Our recognition of all that went right was a slight release of tension. I guess it was an understated acknowledgment that we might yet succeed in keeping everyone on board alive.

Of course, there was no time or inclination to celebrate.

Yes, it was a relief that one of the biggest problems we faced that day had been solved: We had gotten the plane down and brought it to a stop in one piece. But we weren't out of the woods yet. This was not yet a successful outcome.

I sensed that the plane was still intact, even though the moment of impact had been a hard jolt, especially in the back of the plane. I assumed that the passengers were probably OK. I'd later learn that some had their glasses knocked from their faces during the landing. Others hit their heads on the seat backs in front of them. But few passengers were seriously injured on impact. After the plane settled in the water, I heard no screaming or shouting from the cabin. Through the cockpit door, I heard just muffled conversation. I knew that the passengers were likely looking out their windows at the dark green water in the river, feeling stunned.

Seconds after the airplane stopped, Jeff turned to the evacuation checklist. The list is split between the captain and first officer, but the captain's duties—including setting the parking brake—are only useful on land, or if we had working engines. I decided not to waste time on things that would have no benefit to our situation there on the river. Jeff's checklist took him ten or fifteen seconds to complete. He checked that the aircraft was unpressurized and that the engine and APU (auxiliary power unit) fire push buttons were pushed.

As he did that, I opened the cockpit door and stated one word, loudly: "Evacuate!"

In the front of the cabin, by the left and right doors, Donna and Sheila were ready for my order. I hadn't had time to inform them during the descent that we were landing in water. But once they saw where we were, they immediately knew what to do. They changed their commands to "Don life vests; come this way!"

They knew to assess the exits carefully. They had to make sure the plane wasn't on fire on the other side of the door and that there were no jagged metal pieces. They knew not to open a door if that portion of the plane was under water. The good news was that we could tell by the attitude of the plane that the forward doors were above the waterline. And so they opened them.

The slide rafts are supposed to inflate when the doors open. That happened correctly on the right side of the plane. On the left side, however, the raft didn't automatically inflate and had to be deployed manually.

A far more dangerous issue: The back of the plane was quickly filling up with ice-cold river water. We later learned that the bottom of the aft end of the fuselage had been violently torn open by our contact with the water when we landed. A rear exit door had been partially opened, very briefly, and couldn't be closed completely, which also brought water into the cabin. The plane was gradually taking on a more tail-low attitude.

Doreen, stationed in the rear of the aircraft, had a deep gash in her leg, the result of metal that had sliced through the floor from the cargo compartment when the plane hit the water. Though the water level rose quickly, she was able to make her way past floating garbage cans and coffeepots, urging passengers to move forward toward usable exits. After she got into the right-front slide raft—actually an inflatable slide that doubles as a raft—a doctor and a nurse who were passengers put a tourniquet around her leg.

Because the waterline was above the bottom of the aft doors, the emergency slide rafts at the aft doors were useless. That meant we needed to use the two overwing exits, which normally wouldn't be opened when a plane is in water. One passenger struggled to push open an overwing window exit. Another knew the exit needed to be *pulled* into the cabin, and did so. This second passenger had been in the emergency row and, luckily, had the presence of mind to read the instructions after we hit the birds. He knew he might be called upon to act and he prepared himself.

As the evacuation began, passengers seemed understandably tense and serious—some were pretty agitated, hurriedly jumping over seats—but most were orderly. A few later called it "controlled panic."

Since the rear exits were not usable, people were bunching up at the wing exits. There was still room in the rafts up front, so Donna, Sheila, and I kept calling for passengers to come forward. I didn't observe people trying to get their luggage, but I later learned some of them did, against the advice of other passengers. One woman, who had collected her purse and suitcase, would later slip on the wing, which sent her suitcase into the river. Another man held his garment bag while standing on the wing, an unnecessary accessory at a time like that.

Jeff noticed that some people still on the plane were having trouble finding their life vests. The life vests are under the seats, and not easy to spot. Jeff told people where the vests were. Some passengers went out on the wings carrying their seat cushions, because they didn't realize there were life vests available to them.

As passengers exited, Jeff and I, along with some young male passengers, gathered more life vests, jackets, coats, and blankets to hand to people shivering out on the wings. We kept handing them out of the plane, as those who were on the wings and in the life rafts shouted that they needed more. The temperature outside was twenty-one degrees, and the windchill factor was eleven. The water was about thirty-six degrees. Those standing on both wings were in water above their ankles, and eventually, some would be in water almost up to their waists. Eric Stevenson had to kneel for balance because late in the rescue the left wing had lifted out of the water as the plane tilted to the right. Its upper surface was "like an ice rink," he thought.

Flight attendants train to empty a plane of passengers in ninety seconds. That's the FAA certification standard. But doing the training in an airplane hangar, with 150 calm volunteers, is a bit different from attempting it in freezing weather in the middle of the Hudson River.

I was proud of how fast the crew got everyone off the plane. The last passenger left the aircraft about three and a half minutes after the evacuation began, even with the aft exit doors unusable and water entering the aft cabin.

Once the plane emptied I walked down the center aisle, shouting: "Is anyone there? Come forward!"

I walked all the way to the back and then returned to the front. Then I took the same walk again. The second time, the water in the back of the plane was so high that I got wet almost up to my waist. I had to stand on the seats as I made my way back to the bulkhead. The cabin was in good shape. The overhead bins were closed, except for a few in the aft part of the cabin. The seats were all still in place.

When I got back to the front, Sheila was in the slide raft on the right side of the plane with a full load of passengers, but was having difficulty detaching it from the airplane. Standing inside the plane, I lifted the Velcro strip that set them free.

Jeff, Donna, and I were the final three people inside the plane. As I finished that final walk down the aisle, Donna spoke to me in no uncertain terms. "It's time to go!" she said. "We've got to get off this plane!"

"I'm coming," I told her.

As is protocol, I grabbed the emergency locator transmitter (ELT) from the forward part of the cabin and handed it to a passenger in the left-front slide raft. Donna got into that same raft and I went into the cockpit to get my overcoat. I also grabbed the aircraft maintenance logbook. I left everything else behind. I reminded Jeff to get his life vest. I already had mine. I handed my overcoat to a male passenger in the left-front raft who was cold.

After Jeff stepped out, I took one final look down the aisle of the sinking plane. I knew the passengers had all made it out. But I wasn't sure if some of them might have slipped into the near-freezing water. How would I describe my state of mind at that moment, as a captain abandoning his aircraft? I guess I was still busily trying to keep ahead of the situation—anticipating, planning, and checking. There was no time to indulge my own feelings. The 154 people outside the aircraft were my responsibility still, even though I knew that rescuers would be working to pick us all up.

By the time I got into the raft, there were already boats around the airplane. The rafts are designed to accommodate forty-four people, with a maximum overload capacity of fifty-five. But we had fewer than forty people on our raft on the left side of the plane, and it felt pretty crowded. I saw no one crying or sobbing. There was no shouting or screaming. People were relatively calm, though in shock from the enormity of our experience. Though we were packed extremely tightly, no one was pushing. People were just waiting to be rescued, and there wasn't much conversation at all. Everyone was very cold, and we were shivering. Though I was wet from walking in water to the back of the cabin, my recollection is that in our raft the bottom was pretty dry.

It was fortuitous that we landed in the river right around Forty-eighth Street, just as several high-speed catamaran ferries were preparing for the afternoon rush hour. Across the river in New Jersey, at the NY Waterway Port Imperial/Weehawken Ferry Terminal, the boats' captains and deckhands were shocked to see our plane splash into the water. They were riveted by the sight of passengers almost immediately escaping from the plane. And in that instant, without being contacted by authorities and on their own initiative, they quickly headed our way. Fourteen boats ended up assisting us, their crews and passengers doing whatever they could to get us to safety.

Ferries aren't designed as rescue ships, of course, but the deckhands rose to the challenges before them. Many had trained and drilled for such an emergency. Others adapted to the situation and worked by their wits.

The first vessel to reach us, just three minutes and fifty-five seconds after we came to a stop in the water, was the *Thomas Jefferson,* under the command of Captain Vince Lombardi of NY Waterway. He began the rescue of passengers from the right wing. His vessel would eventually rescue fifty-six people, more than any other vessel that day.

The *Moira Smith,* the second vessel to arrive, commanded by Captain Manuel Liba, approached our raft. I shouted to the crew members on that boat, "Rescue people on the wings first!" Passengers on the wings were obviously in a more precarious situation. None of the passengers on our raft objected as the boat turned away from us. People really did seem to grasp the entire scope of the situation, rather than just their individual needs, and I was grateful for their goodwill. Those shivering in our raft clearly understood that the people standing in water on the wings had to be rescued first.

I wanted to get a head count. I knew there were 150 passengers and 5 crew members on the plane. Could we add up those in the rafts and on the wings and see if we'd reach 155?

I asked those on my raft to count: "One, two, three, four . . ."

I then yelled to a man on the left wing to get a count of people on his wing. He tried, but the process was soon overcome by events, and besides, by this time, people were already being rescued and taken off the wings and rafts. I couldn't see the raft and the wing on the other side of the airplane or communicate with the people over there. So we never were able to get any kind of count while still on the river.

Our raft remained tethered to the left side of the airplane and Jeff expressed concern that as the plane continued to take on water and ride lower, it might eventually pull the slide raft down and tip people out into the river. He spent several minutes trying to disconnect us from the plane.

"I can't get it undone!" he said as the plane inched lower in the water. A knife is stored on each raft, but with so many people crammed together, and so much going on, it wasn't immediately evident to us where the knife was. I knew that deckhands on boats usually carry knives. So I shouted to someone on the raft closer to the ferry to call up for a knife. A folding knife was produced, tossed toward our raft (a woman passenger caught it), and Jeff was able to cut us loose.

When passengers were later asked how long they waited for the lifeboats to arrive, some estimated it took fifteen minutes or longer. Actually, the first ferry had arrived in under four minutes. Standing in freezing water, after the trauma of a life-threatening emergency, can alter a person's sense of time. After just a few minutes outside in the water, many of those on the wing were unable to stop shaking. A quick rescue was imperative to minimize hypothermia.

One passenger had jumped into the water and began swimming to the New York side of the river. He soon thought better of it, given the water temperature, and swam back. Other passengers pulled him into our raft, and we saw that he was unable to stop shaking.

One of our passengers was Derek Alter, a first officer for Colgan Air. "Sir, you have to get out of these clothes, and you have to do it now," Derek told the man who had been in the water. Derek took off his first officer's uniform shirt, gave it to the man, and then kept his arm around him to keep him warm. (Derek later said that it was his Boy Scout training that helped him know that the man needed to get out of his wet clothes immediately.)

The third vessel to arrive, the NY Waterway ferry *Yogi Berra*, captained by Vincent LuCante, rescued twenty-four people.

One woman slipped off the wing and into the river, and two other passengers risked falling in themselves as they pulled her back up. When it was time to get her on a ladder, she was unable to move her legs from the cold, and she fell off and had to be helped on again. Others also fell into the water trying to get up the ladders. It was pretty harrowing. Then there was the release of emotions. When passengers finally made it onto the ferries, some of them hugged the deckhands.

One ferry captain was Brittany Catanzaro, just nineteen years old, whose regular job was to transport commuters from Weehawken and Hoboken, New Jersey, to Manhattan. Her ferry, the *Thomas Kean*, the fourth vessel to arrive, had been pointed away from us when we landed, but she turned it around and headed our way. She and her crew members pulled twenty-six passengers off the wings. All the ferries had to be careful and slow down, especially as they approached those who were standing on the wings. If they threw off too big a wake, passengers could have been knocked into the water. Maneuvering near the aircraft was difficult, especially with the strong current, and required great ship handling to prevent bumping the plane.

An NYPD helicopter arrived, and I watched a diver being dropped from it into the river. The downwash from the rotors was strong; spray from the surface of the river got into our eyes. That was cold water mixed with a cold wind. The police diver rescued a passenger in the water near one of the wings.

Jason's Cradles, hammocklike maritime rescue devices with cloth webbing and similar to ladders with rungs, were lowered from the boats to us in the rafts, and passengers started climbing up. At one point, there were fears that the stern of a ferry might puncture a raft, so it had to move away and reposition itself. One elderly female passenger did not have the strength to climb onto the deck of the boat. The hammocklike part of the Jason's Cradle had to be used with pulleys to get her on board.

When it came time for the *Athena,* a Block Island ferry used by NY Waterway and captained by Carlisle Lucas, to rescue those in our raft, I shouted, "Injured and women and children first!" Others on our raft passed the message up to the deckhands. It seemed like we were all on the same page.

I wasn't just being chivalrous. Because women and especially children weigh less than men, they would be more susceptible to hypothermia. They would also lose physical strength more quickly. So it made the most sense to get them onto the boats sooner.

As things turned out, though, it wasn't logistically easy to help the women and children first. Because the raft was so full and movement within it so difficult, those closest to the end of the raft, nearest the ferry, were taken off first.

In the stress of the moment, there was an efficient kind of order that I found absolutely impressive. I also saw examples of humanity and goodwill everywhere I looked. I was so moved when deckhands on ferries took off the shirts, coats, and sweatshirts they were wearing to help warm the passengers.

As a boy, I had been upset by the story of New Yorker Kitty Genovese and the bystanders who had ignored her. Now, as a man, I was seeing dozens of bystanders acting with great compassion and bravery—and a sense of duty. It felt like all of New York and New Jersey was reaching out to warm us.

While we were on the river, Patrick, the controller who had overseen our flight from his post on Long Island, was relieved of his position and invited to go to the union office in the building. He knew, as did his superiors, that he shouldn't finish his shift, guiding airplanes still in the sky. Controllers are always asked to step away from their duties after major incidents.

Patrick was understandably distraught. He assumed we had crashed and that everyone on the plane had perished. "It was the lowest low I had ever felt," he later told me. "I was asking myself: What else could I have done? Was there something different I could have said to you?"

He wanted to talk to his wife but feared he would fall apart if he did. So he sent her a text: "Had a crash. Not OK. Can't talk now." She thought he'd been in a car accident. "Actually, I felt like I'd been hit by a bus," he said. "I had this feeling of shock and disbelief."

Patrick was secluded in that office with a union rep who kept him company and talked him through it. There was no TV, so he couldn't see coverage of the rescue. In case we had a bad outcome, his union rep didn't think Patrick needed to see it in those early moments.

Over and over again, Patrick played in his mind his final exchanges with me, assuming they were my final words. He had heard the distress in pilots' voices during lesser emergencies he'd dealt with in the past. As he would describe it, their voices became "almost like a quiver." He thought about my voice, and how it seemed "strangely calm."

At that point, he didn't know what I looked like and didn't know anything about me. He just knew we had spent a few riveting minutes connected to each other, and now he assumed I was gone.

He was told he couldn't leave the facility until the drug testers came to take a urine sample and do a Breathalyzer test. This is standard procedure for controllers—and pilots, too—involved in an accident. It's part of the investigation.

Patrick sat in that union room, consoled by the union rep, for what felt like hours. Then a friend poked his head into the room and said, "It looks like they're going to make it. They're on the wings of the plane."

Patrick later told me that his relief was beyond words.

One of the passengers was sitting near Jeff and me in the raft. Like so many people, he was drained and emotional. But he wanted me to know that he appreciated what the crew and I had done to bring the plane down safely.

He took my arm. "Thank you," he said.

"You're welcome," I told him.

It was the simplest exchange between two men at an extraordinary moment, but I could tell it meant a great deal to him to say it. It meant a great deal to me to hear his words, and for Jeff and Donna, near us, too.

The cold air and wind were not immediately debilitating. But as we all waited for our turn to be rescued by the ferry *Athena,* a lot of us were in pretty rough shape. Many couldn't stop shivering.

I made sure I was the last person off the raft, just as I had wanted to be the last person off of the plane. I don't think there are any written guidelines suggesting that the captain be the last to leave a plane or any other vessel during an emergency. I was aware of the maritime tradition, but that wasn't the reason I did it. It was just obvious to me: I shouldn't be rescued until all the passengers in my care were attended to.

The rescue went quickly, all things considered. The deck of the ferry was about ten feet above the raft, so it took some effort for passengers to make their way up. By the time it was my turn to climb up the ladder, I was so cold that I could no longer use my hands. I had to stick my forearms through the rungs. I couldn't grasp anything with my fingers.

From the deck of the ferry, standing with seventeen other survivors from Flight 1549, including Jeff, I looked back at the airplane. It continued to slowly sink lower in the water, as it drifted south toward the Statue of Liberty surrounded by a small trail of debris and leaking jet fuel.

Standing there, I realized I still had my cell phone on my belt. Though my pants were drenched, the phone was dry and working. It was my first moment to call Lorrie.

We have two landlines in the house and she has a cell phone, but I couldn't get through to her on any of them. She wasn't answering because she was on one of the lines, talking to a business associate. She saw my number come up on her cell phone, but at first she ignored it.

Given all the ringing, she told the person she was talking to: "Sully is calling every line in the house. Let's see what he wants."

She answered the other line, saying, "Hello."

Hearing her voice, not knowing what she knew or didn't know, my first words were meant to reassure her: "I wanted to call to say I'm OK."

She thought that meant I was on schedule to fly back to San Francisco that night.

"That's good," she told me. She assumed I had already landed Flight 1549 in Charlotte. I saw she needed an explanation.

"No," I said. "There's been an incident."

She still wasn't getting it. She didn't have her TV on, so she was unaware of the nonstop coverage of the incident that was all over the national cable networks. She assumed that I was trying to tell her my flight was delayed, and that I might not make it home.

And so I told her straight, almost as if I was giving her bullet points. "We hit birds. We lost thrust in both engines. I ditched the airplane in the Hudson."

It was a lot for her to digest. She paused and asked her first question. "Are you OK?"

"Yes," I told her.

"*OK* OK?" she asked. Obviously I had survived. She was asking if I was OK in a broader sense.

"Yes," I said. "But I can't talk now. I'm on my way to the pier. I'll call you from there." I felt pretty emotional hearing her voice; I could have used her consoling words. At the same time, there was so much to tell her and no time to do so. I wanted the kids to know I was safe, too. Until I could get back to them, they'd be hearing everything from the news reports on TV. But at least I had made contact.

After my call, Lorrie lay down on the bed in our bedroom. She wasn't crying, but she was shaking really hard. My call had been a shock. She called a close friend and said, "Sully just crashed an airplane and I don't know what to do." Her friend told her, "Go get your girls." So she got the girls out of school and brought them home.

While still on the ferry, I began running through my mental checklist of other things I should be doing. I knew that US Airways was well aware of the incident through Air Traffic Control, but I thought I'd better give the airline a sense of the situation from my end.

Every flight has an airline dispatcher assigned to monitor it. The dispatchers work at their computers in a large, windowless room at the US Airways Operations Control Center in Pittsburgh, and they each track many flights at the same time.

I called Bob Haney, who was on duty that day as US Airways' airline operations manager, and after a few rings he picked up.

"This is Bob," he said. His delivery was clipped, and there was an intensity in his voice.

"This is Captain Sullenberger," I said.

"I can't talk now," he told me. "There's a plane down in the Hudson!"

"I know," I said. "I'm the guy." He was momentarily speechless. He couldn't believe that the pilot from the aircraft in the Hudson, a scene he was watching on TV at that moment, was calling his desk phone. Given the gravity of the situation, we quickly began discussing the matters at hand. But I'd later smile at the memory of how he tried to cut me off at the beginning of our conversation with breaking news. "There's a plane down in the Hudson!" Yes, I knew about that.

The *Athena* docked at Manhattan's Pier 79, let us off, and then went back one more time to the plane to make sure no one was left behind. By 6:15 P.M., it would return to duty shuttling commuters back and forth across the Hudson, its seats still wet from the soaking Flight 1549 survivors.

As soon as I stepped onto the pier at the ferry terminal, I was met by US Airways captain Dan Britt, our union rep at LaGuardia. He had seen the television coverage at his home in New York, put on his uniform, and come down to be with me and Jeff.

I asked him to help me get answers and updates, and we both started making calls, verifying that the injured were being treated. I walked over to Doreen, who was on a gurney and was being treated by an EMT. She was the most seriously injured, with a gash in her leg, and would remain hospitalized for several days. I gathered together the rest of the crew, and included our two other airline pilot passengers, American Airlines first officer Susan O'Donnell and Colgan Air's Derek Alter, who had given his shirt to a passenger in the raft.

Some passengers had been taken to the New Jersey side of the river and the rest came to New York, so it was hard to keep track. I desperately wanted a tally of all those who had been rescued, but I was still unable to get any kind of confirmation. The authorities kept asking me for the manifest. On domestic flights, the crew is not given one. US Airways would spend some time constructing one from the electronic records of the flight.

Police were everywhere, and a high-ranking police officer told me that Mayor Michael Bloomberg and Police Commissioner Raymond Kelly wanted me to go see them at another location. I had to decline. "I have responsibilities here," I said. And so Mayor Bloomberg and Commissioner Kelly ended up coming to the ferry terminal to ask me a few questions. I was too concerned about the passenger issues to have a real conversation with them. I gave them a short update and that was it. "I made sure everyone was off the airplane," I told them. "We're trying to find out if they're all accounted for."

Much discussion took place about where the crew and I should go next. Eventually we were taken to the hospital to be evaluated and have our vital signs checked. All the while I kept asking and asking, "What's the total?"

After we were examined in the emergency room and were told we were all OK, we were left just standing around, waiting for confirmation, waiting for news, waiting to find out where we would go next. There weren't enough chairs for all of us in the examination room, but I didn't feel like sitting anyway. It was stressful, just waiting, not knowing the outcome, standing there in my wet uniform and my wet socks. I wouldn't have a chance to get into anything dry until midnight.

In the hour or two that followed, three more doctors came in. They didn't really have any medical reason for stopping by. They probably were just curious to get a look at us, given that we were all over the news. At one point, a doctor in his mid-forties stopped in and looked me right in the eye. I could tell that he was trying to get the measure of me, trying to figure out what made me tick. He didn't say a word for fifteen or twenty seconds. Finally he spoke. "You're so calm," he said. "It's incredible." He was mistaken. I didn't feel calm at all. At that point, I was feeling numb and out of sorts. I just couldn't relax until I knew the count was 155.

Finally, at 7:40 P.M., more than four hours after we landed in the Hudson, Captain Arnie Gentile, a union rep, came in and gave me the word. "It's official," he said.

I felt the most intense feeling of relief I'd ever felt in my life. I felt like the weight of the universe had been lifted off of my heart. I probably let out a long breath. I'm not sure I smiled. I was too spent to celebrate.

It had been the most harrowing day of my life, but I was incredibly grateful for this ending. We hadn't saved the Airbus 320. That was ruined. But the people on the plane, they would be returning to their families. All of them.

16

STORIES HEARD, LIVES TOUCHED

I am used to it now.

I open a letter and five one-dollar bills fall out. "Mr. Sullenberger, Great job! I'd like to buy you a beer, albeit a cheap domestic one."

A fax arrives: "In this crazy world, it's good to know that chance still favors the prepared mind. Good job, Captain!"

A letter comes with an illustration of Snoopy in an exhilarated dance pose. The caption: "Oh Happy Day!" The letter writer is a woman from New Jersey. "We on the East Coast are still scarred by 9/11. It seemed all in the tristate area lost a family member, a friend, a neighbor, a coworker. Your splash in the river made us feel elated, serene, and happy!"

I have gotten thousands of messages such as these since Flight 1549. I have received ten thousand e-mails from people who tracked down my safety consulting business online. Another five thousand e-mails arrived at my personal e-mail address. I don't know much about Facebook, but my kids tell me I have more than 635,000 fans there.

I've heard from people on every continent except Antarctica. And almost every time I'm at the mall or in a restaurant, strangers come up to say they don't mean to bother me, but they just want to say thank you.

While a few of these correspondents had loved ones or friends on Flight 1549, the vast majority did not. What happened on that airplane touched them deeply enough that they felt compelled to reach out to me and my family. Some tell me that after hearing about our flight, they found themselves reflecting on a seminal moment in their own lives or thinking about a person who inspired them. Others ended up reviewing the dreams they had for their children or feeling renewed grief about losses they're still trying to understand.

I have become a recipient of people's reflections because I am now the public face of an unexpectedly uplifting moment that continues to resonate. Hearing from so many people, paying attention to their stories—that's part of my new job.

I've come to see their thankfulness as a generous gift, and I don't want to diminish their kind words by denying them. Though it made me uncomfortable at first, I've made a decision to graciously accept people's thanks. At the same time I don't strive to take it as my own. I recognize that I have been given a role to play, and maybe some good can happen as a result.

It's not a role that I had ever experienced before. I spent a lifetime being anonymous. I was proud of my wife, proud of my kids, but I lived a quiet home life. My work life was also mostly hidden, conducted on the other side of a locked cockpit door.

But now I am recognized everywhere, and I have people coming up to me with tears in their eyes. They're not sure why they're crying. Their feelings about what the flight represents, and then the surprise of meeting me, just cause a swell of emotions. When people seem so grateful to me, my foremost feeling is that I don't deserve this attention or their effusive thanks. I feel like a bit of an impostor. And yet, I also feel I have an obligation not to disappoint them. I don't want to dismiss their gratitude or suggest that they shouldn't feel the way they do.

Of course, I'm still not comfortable with the "hero" mantle. As Lorrie likes to say, a hero is someone who risks his life running into a burning building. Flight 1549 was different, because it was thrust upon me and my crew. We did our best, we turned to our training, we made good decisions, we didn't give up, we valued every life on that plane—and we had a good outcome. I don't know that "heroic" describes that. It's more that we had a philosophy of life, and we applied it to the things we did that day, and the things we did on a lot of days leading up to it.

As I see it, rather than an act of heroism, that philosophy is what people are responding to.

They also embraced news of Flight 1549 because it came at a moment when a lot of people were feeling pretty low.

On January 15, 2009, the day of our flight, the world was in transition. The presidency of the United States was about to change hands, which had some people feeling hopeful and others feeling nervous about the road ahead. It was a time of great uncertainty, with two wars and the world economy falling apart. On a lot of fronts, people felt confused and fearful. They wondered if we as a society had lost our way or gotten off track. Some people had been questioning even our basic competence.

They heard about Flight 1549 and it was unlike most stories they learn of through the media, in that the news continued to be good. The plane had landed safely. Passengers and rescuers had reached out and helped one another. Everyone on the plane had lived. It was all positive news (unless of course you owned or insured that Airbus A320—then the news wasn't as completely upbeat).

For people watching reports of Flight 1549 on their televisions, this felt remarkable. It enabled them to reassure themselves that all the ideals that we believe in are true, even if they're not always evident. They decided that the American character still exists, that what we think our country stands for is still there.

I've come to have a greater appreciation of life—and of America, too—through my interactions with so many people since the event. They say they were touched by my story, but so very often I am even more touched by theirs.

When Flight 1549 landed in the Hudson, eighty-four-year-old Herman Bomze watched the rescue from his thirtieth-floor Manhattan apartment overlooking the river.

Mr. Bomze, a retired marine and civil engineer, found himself feeling very moved as passengers scurried into their rafts and onto the wings. He was concerned that all the passengers hadn't made it out of the plane. He worried the ferries wouldn't get to everyone in time. He called his daughter, Bracha Nechama, and left her a voice mail to tell her how it affected him. She in turn sent a letter telling me his story.

US Airways Flight 1549 had just taken off from New York's LaGuardia Airport when we struck birds, permanently damaging both engines and forcing an emergency landing in the Hudson River. (*Associated Press*)

A dramatic photographic sequence of Flight 1549 landing in the Hudson River taken from a security camera. (*Associated Press*)

We landed close to the ferry terminals, so the first responders were able to reach the aircraft quickly and rescue the passengers and crew. (*Associated Press*)

Despite my efforts to account for the passengers during the rescue, it was not until hours later that I received the final word: There were no fatalities, and all 155 passengers and crew were safe. (*Associated Press*)

The events of January 15, 2009, would not have ended the way they did without the tremendous work of the first responders. The first ferryboat was at the scene in less than four minutes, ensuring the survival of all the passengers on board. *(Associated Press)*

I will forever be grateful to the first responders for their courage, skill, determination, and quick actions. (*Associated Press*)

The air temperature that day was 21 degrees and the water was 36 degrees. (*Associated Press*)

On Monday, February 9, 2009, New York City mayor Michael Bloomberg honored the crew of Flight 1549 with keys to the city. *(Associated Press)*

The aircraft, still in the Hudson River in the early evening, waiting to be salvaged. *(Associated Press)*

The crew of US Airways Flight 1549 *(from left to right):* Flight Attendant Doreen Welsh, First Officer Jeffrey Skiles, Captain Chesley Sullenberger, Flight Attendants Donna Dent and Sheila Dail. *(Nigel Parry/CPi Syndication)*

I first met Air Traffic Controller Patrick Harten *(center)* on February 24, 2009, when we both testified before the Aviation Subcommittee of the House Transportation and Infrastructure Committee. His quick thinking and single-minded focus were instrumental in helping us achieve a successful outcome on January 15. *(Associated Press)*

On June 22, I visited some of the captains and crew from NY Waterway who participated in the passenger rescue. Standing fifth from the left is Arthur Imperatore, Sr., owner of NY Waterway. The three people in the white uniform shirts are (*from left to right*) Captain Manuel Liba of the *Moira Smith,* Captain Brittany Catanzaro of the *Governor Thomas Keane,* and Captain Vince Lombardi of the *Thomas Jefferson.*
(*Daniel H. Birman*)

The airplane is in storage at J. Supor & Son Trucking and Rigging in New Jersey.
(*Daniel H. Birman*)

On April 15, 2009, I was honored by the United States Air Force Academy with the Jabara Award for Airmanship. My classmate, Superintendent Lt. Gen. John Regni, presented the award.
(Air Force Photo/Mike Kaplan, DenMar Services Inc.)

I have had many remarkable experiences since January 15, but few are as memorable as the reception I received upon returning to my alma mater.
(Air Force Photo/Dave Ahlschwede, DenMar Services Inc.)

The Academy arranged for me to take a flight in a glider during my visit. *(Air Force Photo/Dave Ahlschwede, DenMar Services Inc.)*

The next generation of Academy cadets is a truly exceptional group of young men and women whose dedication and service to their country is inspiring. As a token of their appreciation, Cadet Wing Commander C1C Jonathan Yates presented me with a hand-carved falcon, the mascot of the Academy. *(Air Force Photo/Dave Ahlschwede, DenMar Services Inc.)*

Since January 15, I have thrown a first pitch for the San Francisco Giants, the Oakland Athletics, and the New York Yankees. It took some practice, but I managed to get the ball to the catcher all three times. Here Lorrie and the girls enjoy our visit to AT&T Park in San Francisco.
(*Alex Clemens*)

Lorrie and Diane Sawyer before our taping of *Good Morning America*, February 9, 2009. (*Alex Clemens*)

Life since January 15 has been a constant adventure for Lorrie and me, taking us to the White House, Buckingham Palace, even an Academy Awards party. Here we are in Tampa, Florida, for Super Bowl XLIII.
(Alex Clemens)

The National Football League honored the crew of Flight 1549 during the pregame celebration before Super Bowl XLIII on February 1, 2009.
(Alex Clemens)

Lorrie and I have always been strong supporters of St. Jude Children's Research Hospital, and I am grateful that I can now play a more significant role in championing their work. I visited with Darcy and her mother, Cathy, on May 29, 2009.
(St. Jude Children's Research Hospital, BMC, Ann-Margaret Hedges)

Upon my return to California from New York, my hometown of Danville hosted a celebration in my honor. I was overwhelmed and touched by the number of my friends and neighbors who came to welcome me home. *(top: San Ramon Valley Fire Department; bottom: Danville Police Department)*

The Sullenberger family with President Obama and the First Lady at an inaugural ball, January 20, 2009. (*Author's Collection*)

I will forever feel a deep connection to the people of New York and New Jersey who behaved so admirably on January 15 and who welcome me so warmly each time I return. (*Alex Clemens*)

In 1939, when Herman was fifteen years old, he, his sister, and his parents were living in Vienna and trying desperately to get out of Austria. Because they were Jewish, their apartment had been ransacked by Nazis. They knew of the mass deportations of Jews and had heard the rumors of mass murder.

Herman's family hoped to come to the United States, where relatives lived and were willing to sign paperwork vouching for them. In those days, the United States had strict quotas on how many European refugees could be admitted. At the U.S. embassy in Vienna, the family was told that only three visas were available—for Herman, his mother, and his sister. Because Herman's father had a Polish passport, and there were different quotas for Poles, there would be no visa for him.

"Please," Herman's mother pleaded. "Let our family stay together."

"You can stay together if you'd like," the embassy clerk told them. "If you want to stay here in Austria, you can be together. If the three of you want to go, you can go. It's your choice."

The family made a decision. Herman's father would stay behind. Herman, his sister, and his mother would escape to the United States, where life would be safer for them. The three of them arrived here in August 1939, and not long after that, Herman's father was transported to Buchenwald concentration camp. He was murdered there in February 1940.

Almost seventy years later, Herman watched the rescue of Flight 1549 unfold, and it was, in part, these difficult memories that compelled him to call his daughter, Bracha. Afterward, Bracha continued to think about the connections between me and her father, and she reached out to me with her letter.

She wrote of Herman's great reverence for life, forged through the Holocaust. She also wrote that her father was lucky that our flight found safety in the river, as opposed to crashing into buildings in Manhattan.

"Had you not been so skilled and such a lover of life," she wrote, "my father or others like him, in their sky-high buildings, could have perished along with your passengers. As a Holocaust survivor, my father taught me that to save a life is to save the world."

She explained to me the Jewish view that if you save one person, you never know what he or she might go on to accomplish, or how his or her progeny might contribute to peace and healing in the world. "May you know the joy of having saved generations of people," Bracha wrote, "allowing them the possibility of humanitarianism such as yours. Bless you, Captain Sullenberger."

Her letter continues to move and inspire me. I feel honored that she viewed the landing of the plane in the Hudson as "a powerful commitment to life." She's right: I don't know the good things still to be accomplished by the 154 people on my flight. I can't fathom what contributions might be made to the world by their children, grandchildren, and great-grandchildren yet to be born.

There were those who wrote to say they agreed with me: I am not a hero. I appreciated the ways they spoke to me. They wrote to say that preparation and diligence are not the same as heroism.

"In your interviews, you seemed uncomfortable being called a hero," wrote Paul Kellen of Medford, Massachusetts. "I also found the title inappropriate. I see a hero as electing to enter a dangerous situation for a higher purpose, and you were not given a choice. That is not to say you are not a man of virtue, but I see your virtue arising from your choices at other times. It is clear you take your professional responsibilities seriously. It is clear that many of the choices in your life prepared you for that moment when your engines failed.

"There are people among us who are ethical, responsible, and diligent. I think there are many of them.

You might have toiled in obscurity were it not for an ill-timed meeting with a flock of birds.

"I hope your story encourages those many others who toil in obscurity to know that their reward is simple—they will be ready if the test comes. I do not mean to diminish your achievement. I just want to point out that when the challenge sounded, you had thoroughly prepared yourself. I hope your story encourages others to imitation."

I heard from more than a few people who lost loved ones in accidents, or who survived accidents themselves. Some of these tragedies involved airplanes.

People wrote of how they had found the courage to return to flying, mostly because they had resolved to trust the professionals in the cockpit.

Karen Kaiser Clark of St. Paul, Minnesota, wrote to me about Delta Air Lines Flight 191, which crashed in Dallas on August 2, 1985, "taking the lives of 139 people, each with a family, a circle of friends, and a place in the world no one could replace. It was wind shear, and my mom, Kate, was among the last seven identified. Her fifteen friends were also killed. Just five months prior, we had taken our father off life support. This was her first venture as a widow."

In the wake of the tragedy, Karen said, she was able to find a path to acceptance and a new appreciation for life. "Following Mom's funeral in Florida," she wrote, "we flew with her ashes and Dad's to inter them in Toledo, Ohio. However, our flight was caught in horrible turbulence. We were all terrified, but in those moments I vowed that if we were able to land, I would find a way to (1) grow through these terrible times and not become bitter, and (2) continue to fly, as I lecture internationally."

Bart Simon, who owns a hair-products company in Cleveland, told me he was on USAir Flight 405 when it attempted takeoff from LaGuardia on the night of March 22, 1992, and crashed in Flushing Bay. "I was one of the lucky ones who walked away with just a small cut on my head," Bart wrote. Twenty-seven people died, and nine of the twenty-three other survivors had serious injuries. The National Transportation Safety Board later said the probable causes were ice on the wings, failures of the FAA and the airline industry to have appropriate procedures regarding icing and delays, and the flight crew's decision to take off without knowing for sure that the wings were free of ice.

"I had been successful in putting that evening out of my mind and getting on with my life," Bart told me, "but the pictures of your landing last month and the similarity of the circumstances—US Airways, LaGuardia, the water—brought the memories rushing back." He wrote that when he watched our crew on TV, it seemed as if we epitomized what passengers hope to find when they board flights: professionals who are "cool, calm, and most of all, in command, no matter how dire the circumstances." He said he was writing to say thanks "on behalf of the millions of us who entrust our lives to you and your fellow pilots every year."

He had boarded a plane out of LaGuardia bound for Cleveland the very morning after that 1992 crash. "The charred remains of Flight 405 were clearly visible in Flushing Bay as my plane taxied by, but I left that morning calm in the knowledge that a skilled professional was at the controls, and that in a short time I would be back home."

As pilots, we sometimes sense that passengers have no awareness of us. It's as if they're just pushing their way past the cockpit, looking for space in the overhead compartments. But in the wake of Flight 1549, I've been able to hear from people such as Karen Kaiser Clark and Bart Simon, and it is humbling to contemplate the faith and trust that they and others like them have placed in us.

Theresa Hunsicker, who runs a day-care center in Louisiana, learned about Flight 1549 while watching Fox News. A forty-three-year-old mother of a nine-year-old girl, she saw me on *60 Minutes* and felt compelled to write about how my interview had affected her.

"My name is Theresa Hunsicker," her letter began, "and I am the daughter of Richard Hazen, who was the copilot of ValuJet 592. It went down in the Florida Everglades on May 11, 1996, with 110 people on board."

Flight 592 had taken off from Miami International Airport, headed for Atlanta, with Captain Candalyn Kubeck at the controls. About six minutes into the flight, she and First Officer Hazen reported fire in the plane's interior and smoke in the cockpit. On the cockpit recording, a female voice is heard shouting from the cabin: "Fire, fire, fire, fire!"

First Officer Hazen radioed to the controller, asking to return to the airport. A few minutes later, traveling at five hundred miles per hour, it crashed in the Everglades. The plane was destroyed on impact.

An investigation revealed that the jet was carrying chemical oxygen generators in its cargo compartment, which likely started or fueled the fire. The oxygen generators had been labeled "empty," and did not have protective shipping caps that could have prevented the fire. The legacy of Flight 592 is that smoke detectors and fire-extinguishing systems are now placed in cargo holds, and changes have been made in how hazardous materials are transported.

In her letter to me, Theresa wrote that she cried watching news reports of Flight 1549. She was reminded of how much she had wished that her father's flight could have had the same positive outcome—a safe water landing. She wished that he and the 109 others on his DC-9-32 could have made their way onto the plane's wings, or into slide rafts in the water of the Everglades.

"I had wondered for many years what my dad's final minutes were like," Theresa wrote. "I had assumed he was full of fear, and regret that he would never see his family again. The thought of him dying in a moment of panic and sadness was overwhelming for me."

Greg Feith, the lead investigator with the National Transportation Safety Board, had told her that her dad's focus would have been on landing the plane. The investigator's words had been somewhat reassuring to her. But in the thirteen years since, she was unable to fully embrace them, because the investigator had never been in a cockpit of a plane in great distress. How could he know what a pilot was truly thinking in such a horrible moment?

That's why my appearance on *60 Minutes* was so meaningful to Theresa. She heard me explain that I had no extraneous thoughts once we lost those engines over

New York. My mind never wandered. I was thinking only of how Jeff and I could get Flight 1549 to safety. My comments provided her with an epiphany of sorts.

"To hear you say how focused you were, and that you had a job to do . . . it gives me peace of mind, because you were someone who lived through it," she wrote. "I now know that Greg was right. My dad didn't leave this world in a moment of deep sadness. He was only trying to do his job. I can't thank you enough, Captain Sullenberger. It has been a real blessing to hear your story."

Lorrie was moved to tears by Theresa's letter. She couldn't get it out of her head, and so she decided to call her. They spoke for an hour—a pilot's wife and a pilot's daughter, sharing memories. "It was cathartic for both of us," Lorrie later told me.

Theresa talked of the inappropriate things well-meaning people have said to her. "People tell me that my father died doing what he loved," she told Lorrie. "Hearing that hasn't been helpful to me. If he died in his garden of a heart attack, that would be different. That would have been dying doing something he loved. But he died in a three-thousand-degree fire. That wasn't what he loved."

The search for the remains of Flight 592 victims took two months, and Theresa told Lorrie how traumatic that was for surviving families. The plane had disintegrated into the smallest pieces, which had to be pulled from the muck far into the Everglades. While workers pushed through every sawgrass blade, snipers stood by to shoot alligators before they approached.

Half of those who died on the flight were never identified. Theresa recalled talking to a woman who was given her son's ankle. They were able to identify it because of a tattoo.

Theresa's father was identified only by a finger, which was delivered to the family in a small box. Because he was in the Air Force, there were records of his fingerprints. "The coroner asked what we wanted to do with it," Theresa said. "We told him, 'We want it back in the Everglades with the rest of him.'"

A mental health counselor and a wildlife and fisheries agent went with the family to the crash site during a memorial service, dropping First Officer Hazen's remains from a small envelope back into the water. It was a surreal and tough moment for the family, and yet it offered a small bit of comfort.

There have been all sorts of airline incidents since the ValuJet crash in 1996, but Theresa said Flight 1549 struck her in ways that none of the others had. Flight 1549 and Flight 592 were similar, she said. Both encountered a serious problem minutes after takeoff. Both couldn't make it back to a runway. Both ended up in the water.

Theresa has been offered the opportunity to listen to the cockpit voice recordings, but has declined to do so. A father of a flight attendant chose to listen, and said he ended up in therapy as a result. The cockpit door was open, and the sounds of screaming passengers are very clear on the tape. "It would be too hard for me to hear that," Theresa said.

In 2006, on the tenth anniversary of the crash, she did find the courage to approach Greg Feith, the investigator: "I can take it," she said to him. "Please tell me: Was my father screaming?" He responded: "Absolutely not. Your dad was going through his checklist. He and Captain Kubeck did everything they were supposed to do until they were incapacitated."

Theresa told Lorrie that when she watched me on *60 Minutes*, "I thought to myself, 'I wish that was my dad. I wish he could have had the same success, and that everyone would be safe, and that it would be him being the hero and giving interviews.' "

She also told Lorrie this: "Because I lived through the worst outcome, I think I celebrate Flight 1549 so much more. My joy for the passengers and crew is so much more profound."

In her letter to me, Theresa explained that she had spent a lot of time over the years thinking about "what-might-have-beens" involving her dad, who was fifty-three years old when he died. He passed away four years before Theresa's daughter, Peyton, was born. "That's the hardest part of the loss," she wrote, "that he'll never meet his granddaughter."

Along with her letter, Theresa enclosed a photo of herself with her husband and daughter—"so you can see who you've touched." They're a very attractive family, pressed tightly together, all smiles. She told Lorrie that she now feels her father and I are connected; two pilots who tried their best to save lives. Though her father would never see his granddaughter, it gave her comfort to know that I would.

And so I was honored to hold the photo of beautiful nine-year-old Peyton in my hands as I thought about First Officer Hazen and the things he has missed.

17

A WILD RIDE

In the early days after Flight 1549, I could sleep only a couple hours at a time. I kept questioning myself. On the very first night, I had said to Lorrie: "I hope they know I did the best I could." That thought remained in my head.

It took me a couple of months to process what had happened and to work through the post-traumatic stress. Our pilots' union has a volunteer Critical Incident Response Program team that began helping me and the crew the day after our Hudson landing. I had asked them for a road map of what to expect. They told me I'd be sleeping less, I'd have distracted thinking, I'd lose my appetite, I'd have flashbacks, and I'd do a lot of second-guessing and "what-iffing."

They were right on all fronts. For the first couple of weeks, I couldn't read a book or newspaper for more than a few seconds without drifting off into thoughts of Flight 1549.

"You might find it hard to shut off your brain," I was told, and that described exactly what I was going through. I'd wake up in the middle of the night and my brain was running hard: What could I have done differently? What did other pilots think of what I had done? Could I have found time to tell the flight attendants that we'd be landing in water? Why didn't I say "Brace for water landing!" when I finally got on the public address system? Could I have done something else, something better?

Eventually, I dealt with the issues in my psyche and started sleeping again. I went through every scenario. For instance, if I had said "Brace for water landing," passengers might have begun fumbling around, desperately searching for life vests, rather than bracing. They might have panicked. The investigation would later show that before we took off, only 12 of 150 passengers had read the safety card in the seat pocket in front of them.

In the end, I was buoyed by the fact that investigators determined that Jeff and I made appropriate choices at every step. But even after I felt comfortable with the correctness of my decisions on January 15, I longed for my life before that day.

For months, if I could have clicked my heels and made the whole incident go away, I would have done so. Lorrie and the girls also wished it had never happened. Though I never thought I was going to die, they certainly felt as if they had almost lost me on January 15. It was hard for them to shake the horror of that feeling.

In time, however, my family came to see that our new reality was manageable, and we tried hard to find the positive possibilities in our new lives. I've been asked by colleagues to be a public advocate for the piloting profession and for airline safety, and I believe that's a high calling. In testimony before Congress, I was able to speak honestly and bluntly about important issues in the airline industry. I know I now have the potential for greater influence in aviation issues, and I plan to be judicious in how I wield that influence.

Meanwhile, the notoriety I gained from Flight 1549 has allowed my family to have more than a few memorable experiences and interactions that otherwise would have been beyond our reach.

We've been plucked from obscurity, and every day the phone rings with an invitation to some new adventure: Buckingham Palace, a Jonas Brothers concert, dinner parties with hosts who would never have noticed us in our previous lives. We're getting used to it, but Lorrie and I still find ourselves looking at each other and saying, "How did we get here?"

Our lives became pretty surreal within minutes of the world's learning about Flight 1549 on that Thursday afternoon.

My uniform was still wet from the Hudson when Lorrie and I began hearing from dignitaries, politicians, and the biggest names in the news media. It wasn't just producers calling, but the on-air personalities themselves: Diane Sawyer, Katie Couric, Matt Lauer. While I was sloshing around the ferry terminal in my waterlogged shoes, back at my house, our two phone lines, the fax line, and Lorrie's cell were all ringing simultaneously. One newspaper reporter even got hold of my daughter Kate's cell-phone number and called looking for me.

By the morning after the incident, while I was still sequestered in New York, dozens of reporters and satellite trucks had gathered outside our house in Danville. Some of them would remain there for ten days.

Lorrie was poised but understandably emotional when she and the girls went outside on Friday morning to give the media a comment. "We've been asked—now I'm going to cry. I have been crying the whole time," she said, then began again. "We have been asked not to say anything by US Air, so we're not going to make any statements about much. But we'd just like to say that we are very grateful that everyone is off the plane safely. That was really what my husband asked to convey to everyone."

A reporter asked how I was faring, and Lorrie answered: "He is feeling better today. You know, he's a pilot. He's very controlled and very professional . . . I have said for a long time that he's a pilot's pilot, and he loves the art of the airplane."

The media picked up on that description, including it in hundreds of stories that followed. Friends and strangers told me that Lorrie wasn't just a beautiful and loving wife. In the emotions of the moment, she turned out to be a pretty good spokesperson, too.

Lorrie was also asked how the family was taking the growing talk that I was a national hero. "It's a little weird—overwhelming," she answered. "I mean, the girls went to sleep last night talking, and I could hear them in the bedroom saying, 'Is this weird or what?' "

I wasn't able to see coverage of Lorrie's impromptu press conference outside our house. In fact, I was too busy to watch any of the media coverage.

The night of the landing, I had gotten just two hours' sleep. There was so much to do that night and the next day. I needed to have my wits about me for interviews with the National Transportation Safety Board. They had a great many questions. How much sleep had I gotten on Wednesday night? What did I eat for breakfast, lunch, and dinner? Was my blood sugar low? How did I feel on my flight earlier in the day? Was I tired? Distracted? How many days earlier was my last drink of liquor? It had been more than a week. It was a beer.

There were a few lighter moments, too. When we got to the hotel on the night of the incident, we were still in our wet clothes. All our belongings, of course, were on the plane. A fellow pilot who had come to help us ran out to a convenience store and purchased toiletries for us. Because we had no dry clothing, he also bought Jeff and me an identical wardrobe: black sweatsuits, black socks, and black, size-34 low-rise briefs. A week later I told him, "My wife liked those low-rise briefs. They're sexier than the whitey-tighties I normally wear." Jeff responded: "Your wife may like yours, but I'm a lot thicker around the middle than you are. Looks like they gave us the same-size briefs. On me, it looks like a thong."

I was in meetings all day Friday, feeling very stressed. I was used up. I was still trying to process everything, and I wanted to clearly recall what happened in the cockpit so I could help investigators sort out the details.

Then I heard that President George W. Bush, with just five days left in office, wanted to talk to me. Next thing I knew, he had called the cell phone of the vice president of our pilots' union, Mike Cleary, who had been by my side for the past twenty hours. Mike handed the phone to me.

"Captain Sullenberger?"

"Yes, Mr. President," I said.

He was very friendly from the start. "You know," he said, "Laura and the staff and I were having something to eat and we were talking about you. I am in awe of your flying ability."

I thanked him. He then had an important question for me.

"Aren't you from Texas?"

"Yes, Mr. President," I said.

He answered like a true Texan: "Well, that explains it!"

I had to smile.

Then he had another question: "Didn't you fly fighters?"

"Yes," I told him. "F-4 Phantoms."

"I thought so," he said. "I could tell."

I didn't ask him how exactly he could tell, but I enjoyed his easy manner, and his Texas-centric view of the whole incident. It was just a pleasant, friendly conversation, and I made sure to tell him that the flight and the rescue were a team effort. I mentioned Jeff, Donna, Sheila, Doreen, the ferry crews, and he acknowledged them.

Despite all that had happened out on the Hudson the previous night, I hung up the phone and just marveled at the way things work in America. Twenty hours before, I was just an anonymous pilot hoping to finish my last flight of a four-day trip, before quietly heading home. Now there I was, talking to the president like we were old buddies from Texas.

About ninety minutes later, I got another call. It was President-elect Barack Obama. He was also very friendly, though a bit more formal in his comments and questions. He invited me to the inauguration, and I immediately knew what my response had to be. I said, "Mr. President-elect, I'm honored, but may I presume to ask that should I be able to attend, it be on the condition that my entire crew and their families accompany me?"

He said yes.

And so we all went, and ended up meeting the new president privately at one of the inaugural balls. Even though it was his big night, he was very gracious and generous in his time with us. He joked with Lorrie. "You're not letting all of this go to your husband's head, are you?" he asked.

Lorrie answered: "People may think he's a hero, but he still snores."

President Obama started laughing. "You've got to tell my wife this," he said. "That's what she says about me." Mrs. Obama was about ten feet away, and he called over to her, "Hey, Michelle, come here, you've got to hear this!"

He had Lorrie repeat her story about my snoring habits, and the two women had a nice laugh at the expense of the president and the pilot.

We kept receiving invitations in the wake of Flight 1549, and some of them we accepted because, well, these would be experiences of a lifetime. How could we turn them down? The Flight 1549 crew was introduced at the Super Bowl, and we got to see the game from perfect seats. Lorrie and I went to an Academy Awards party, where she sat next to Michael Douglas and I got to talk at length with Sidney Poitier.

I was invited to throw out the first pitch at the second game held at the new Yankee Stadium. I made sure I was prepared—I didn't want to embarrass myself in front of fifty-two thousand Yankee fans—so I practiced for the pitch a few days a week for more than a month at a baseball diamond near my house. One of my neighbors, Paul Zuvella, a former major-league infielder who played with four teams, including the Yankees, was kind enough to coach me. I thought I was doing OK, but when it came time for my big pitch, it was a little outside. At least it didn't bounce. On the West Coast, I was also asked to throw first pitches at a San Francisco Giants game and an Oakland A's game.

Though I got the most attention, being the captain of the flight, I was pleased when Jeff, Donna, Sheila, and Doreen were recognized for everything they did. They were at first reluctant to enter the media spotlight, but then they realized that they could help give insights to the world about what it takes to work in the airline industry. Jeff had his share of perks—he got to throw out the first pitch at the Milwaukee Brewers' home opener—and he carried himself incredibly well in interviews. People also got to see that our three flight attendants were highly experienced and well trained; they helped save lives on January 15. Their story reminded everyone that flight attendants aren't just on board to serve coffee and peanuts. They're on the front lines with passengers, ensuring their safety, while we pilots are locked behind closed doors. Despite their initial reticence, Doreen, Sheila, and Donna came to feel an obligation to their peers to be as effective as they could as spokeswomen for their profession. They were class acts all the way. I was very proud of them.

There was a lovely welcome-home ceremony in my hometown of Danville, attended by two thousand resi-

dents. Later, I was invited to speak at graduation cere-
monies back at my alma mater in Texas, Denison High
School. I was beyond thrilled to see ninety-one-year-
old Evelyn Cook, the widow of L. T. Cook Jr., who
had taught me to fly from his grass strip. What a great
honor it was to publicly recognize Mr. Cook's influence
in my life, and to do so before such a large hometown
crowd. It was also fun to be able to say, in front of the
governor of Texas, former classmates, and the town's
dignitaries: "How come you weren't this nice to me
back in high school?"

Had even one person died on Flight 1549, I don't
think I would have accepted any of these invitations.
The whole incident would have had a much more
somber feel to it. But the fact that all of us on the plane
had lived made people want to celebrate, and I saw
that participating in these events was meaningful to
people—and to me.

It also became possible to laugh about the flight.
Comic Steve Martin went on *The Late Show with David
Letterman* and claimed to have been on board with us.
Letterman then showed alleged footage of Steve Martin
walking on the wings, pushing other passengers into
the Hudson, so he could get to the VIP rescue boat. His
little performance was very funny, even for those of us
who had lived through it.

I was amused when businesses began taking advantage of the hoopla over the flight. Several entrepreneurs printed up "Sully Is My Flyboy" baseball caps and "Sully Is My Copilot" T-shirts, and one explained that he did so "because the flight was a sign that good things still happen in the world." The T-shirts were a bit embarrassing for me, but I was OK with them. And in any case, my actual copilot, Lorrie, was always there to keep things from going to my head.

One day in Los Angeles, we got into an elevator where people recognized me. When we got off, a young woman pulled out her cell phone and could be heard telling a friend: "It's so cool! I just ran into Sully the pilot!"

As she talked excitedly on the phone about meeting me, Lorrie was just ahead of her and couldn't help turning around at the mention of my name.

The young woman thought Lorrie had been just another random person on the elevator. "Wasn't that the coolest thing, bumping into Sully like that?" she said.

Lorrie answered, "Well, I'm his wife."

The young woman was a bit embarrassed. "Oh, I'm sorry. It's just that Sully's story makes everyone feel so good. What he did on that flight was so impressive!"

Lorrie smiled, and reassured her that I'm a regular guy—and not always so impressive. "Listen," she said,

"I saw him walking around the hotel room this morning in his underwear."

The woman walked off, talking into her cell phone. I'm guessing she told her friend all about Lorrie's report from our hotel room.

In the weeks after Flight 1549, I finally got to read some of the newspaper stories and see a bit of the TV coverage. For the most part, the media did a pretty good job.

There was an incorrect description of me in one newspaper story that ended up getting repeated around the world. A "police source" was quoted as saying: "After the crash, Mr. Sullenberger was sitting in the ferry terminal wearing his hat, sipping his coffee and acting like nothing happened." A rescuer was quoted as saying: "He looked absolutely immaculate. He looked like David Niven in a pilot's uniform—he looked unruffled. His uniform was sharp."

Yes, I was in uniform, but wearing a hat is now optional for pilots at my airline. It hasn't been required for years, and I'm not big on wearing the hat. In fact, on January 15, my hat was at home in my bedroom closet in California. I also would argue with the dapper David Niven reference. I was actually feeling wet, rumpled, and a bit shell-shocked. (I did appreciate the comparison to David Niven, however, especially given his World War II service during the invasion of Normandy.)

Because of the great interest from journalists—the week after the flight, we were getting 350 media requests a day—I eventually agreed to do a few interviews. I wasn't especially comfortable on TV. I'm still not. It doesn't feel natural to me. But I feel I've gotten the hang of it now.

As things turned out, despite my initial unease before the cameras, I've done OK. There are a great many things I don't know, but there are things I'm pretty sure about, including a lot of issues related to aviation. Most of what the media have asked me about are things I know, so I didn't feel constantly stumped.

I also decided early on that I shouldn't obsess or worry about the media, because they're asking me about me, and of course, I know more about me than anyone else. I was rarely asked questions that were especially technical, and I made a point not to use too much jargon.

Many publications asked to conduct the first print interview with me, and rather than choose between, say, the *Wall Street Journal,* the *Washington Post,* and the *New York Times,* I decided it would be fun if I just went with the *Wildcat Tribune.* That's the student newspaper at Dougherty Valley High School, which Kate attends. Jega Sanmugam, a sophomore and the front-page editor, did the interview. He was prepared. He was sharp. He asked great questions. And he didn't make me nervous.

I also liked the idea of appearing in a newspaper that Kate actually reads. If I showed up in the *Wildcat Tribune*, maybe she would even think I was kinda cool.

While in New York for some interviews, Lorrie, the girls, and I took a break and went to see *South Pacific* at Lincoln Center. As we sat in the audience during the curtain call, the female lead, Kelli O'Hara, spoke about Flight 1549 and mentioned that I was in the audience. The spotlight focused on the four of us, and we then received a ninety-second standing ovation from our fellow theatergoers, which left Lorrie in tears. It was a graphic illustration to her of the enormity of the story of Flight 1549.

She was most moved because she sensed that they weren't just standing for me and for the crew. As she saw it, they rose for that ovation because the success of Flight 1549 had given them a positive sense of life's possibilities, especially in tough times.

People had been losing their jobs in large numbers. Home foreclosures were up. Life savings had been decimated. A lot of people felt like they had been hit by a double bird strike in their own lives. But Flight 1549 had shown people that there are always further actions you can take. There are ways out of the tightest spots. We as individuals, and as a society, can find them.

So at that performance of *South Pacific*, Lorrie thought the audience was standing as a tribute not to Flight 1549, but to what it represented. It represented hope.

I waved at the crowd while Lorrie dabbed at her eyes. Then I hugged her and waved again.

Not long after the Hudson landing, Jeff, Doreen, Donna, Sheila, and I met with dozens of Flight 1549 passengers and their families at a reunion in Charlotte. It was, as you can imagine, a day filled with great emotion for all who were there—the crew, the passengers, and the family members who accompanied them. "Thank you for not making me a widow," one woman told me. Another said: "Thank you for allowing my three-year-old son to have a father." And a young woman who had been on the plane came up to me and said, "Now I get to have children."

Some passengers took the time to introduce me to everyone they had brought with them. "This is my mother, this is my father, this is my brother, my sister . . ."

It went on like that for close to two hours.

In the abstract, 155 is just a number. But looking into the faces of all of those passengers—and then the faces of all their loved ones—it brought home to me how profoundly wonderful it was that we had such a good outcome on Flight 1549.

At the end of the reunion, I thanked them all for coming. "I think today was as much, and as good, for me and my crew as it was for you," I said. "We will be joined forever because of the events of January fifteenth, in our hearts and in our minds."

I had received a letter a few days earlier from a passenger named David Sontag. A seventy-four-year-old writer, film producer, and former studio executive, David is now a professor in the department of communication studies at the University of North Carolina at Chapel Hill. He was on Flight 1549 returning from his older brother's funeral. From his seat, 23F, he saw flames in the engine. He decided to say a prayer as we descended: "God, my family does not need two deaths in one week."

He sent letters thanking me and the crew, and shared words that he had delivered at his brother's memorial service: "We leave a little bit of ourselves with everybody we come in contact with." He also told me that the crew would live on "as a part of all of us who were on board the flight—and everybody we touch with our lives."

I was humbled to think of the connections I now have to each passenger on that plane, to their spouses and their children. It was my honor to spend time with all of them.

So many people came into my life because of Flight 1549—ferry-boat captains, police officers, investigators, journalists, bystanders, witnesses.

Again and again, I return to thoughts of Herman Bomze, the eighty-four-year-old Holocaust survivor who sat in his high-rise overlooking the Hudson River, believing in his heart that saving one life can save the world. And then I think of those on the plane itself; passengers, such as David Sontag, who have now vowed to wrap that lovely thought into the rest of their lives.

David's letter to me was haunting and moving, and I later wrote back to thank him for his kind words. I told him: "As I will live on in your life, you will live on in mine."

18

HOME

It's true for all of us.

Everyone we've ever known and loved, every experience we've had, every decision we've made, every regret we have had to deal with and accept—these are what make us who we are. I've known this all my adult life. Living through Flight 1549 has only reinforced my understanding of what defines our lives.

In the wake of that flight, I have thought about all of my major relationships—my mom, my dad, my sister, Lorrie, the kids, close friends, colleagues.

My father, especially, remains in my mind.

I learned a great many things from him about the importance of being a man of your word, about serving your community, about valuing family and the precious time spent with your children. I smile at my warmest memories of him, including those days when he would close down his dental office for the day so he could lead us on a hooky-playing pirate adventure in Dallas.

I am grateful for the faith he had in me. From the time I was about twelve years old, he'd let me take a rifle and go out in the woods for target practice. He knew the best way to learn responsibility was to be given the opportunity to be responsible, and at as young an age as possible.

In his own life, my dad was content on a lot of fronts. He was content with his modest income, content with living a provincial life in Texas, content with a house that was far from perfect but pleased him because we built it with our own hands. I think of my father when I hear Sheryl Crow sing "Soak Up the Sun." He lived a line from that song: "It's not having what you want/It's wanting what you've got."

But there are darker memories, too, when I think of my father. He wouldn't talk much about his depression—what he lightly called his "blue funk"—and my family never knew the depths to which his inner demons took him.

In the mid-1990s, my father began having gallbladder problems, but he didn't go to the doctor until the pain was fairly acute. Then his gallbladder burst and he needed surgery. He spent weeks in intensive care and was put on a strong course of antibiotics. Some of his organs began to fail. My dad was in pain, and he knew it would take many months to regain his strength, but he was expected to make a complete recovery.

When he was finally sent home from the hospital on December 7, 1995, my mom got him settled in their bedroom. Then she went into the kitchen at the other end of the house to get him some juice, leaving him alone in their room. She heard a noise, a muffled pop. She thought she might have recognized the noise, and then she thought she knew exactly what it was. She dropped the glass of juice, letting it shatter on the floor, and ran across the house back toward the bedroom.

As she was running, she was hoping and wishing that she was wrong about that noise. She entered the bedroom, shouting, "Oh no! Oh no!" It was too late.

My dad had shot himself with a handgun.

He was seventy-eight years old, and he had given no indication that he was planning to do this. He left no note.

It was so distressing that my mom had to be the one who found him and called 911. She had to be the one who washed the bedspread, who got the stain out of the carpet, who called the handyman to fix the glass which the bullet had cracked.

I can't begin to fathom my father's pain, or why he made the decision he did. I assume that like so many suffering from depression, he couldn't help but become inwardly focused. His view of the world was skewed and he probably had tunnel vision, seeing only his problems, unable to have a wider perspective. I think my father just felt so much psychic pain that he couldn't stand it.

He may have believed that he was protecting my mother from having to look after an aging man who likely would need long-term care. Maybe he thought he was acting nobly by saving her from that responsibility. He was also a proud man. It was hard for him to imagine not being self-sufficient.

At the time of his suicide, I was forty-three years old. Naturally, I was distraught, angry, and upset with myself. I thought that I should have been paying closer attention to him. Intellectually, my mom, my sister, and I knew better. As with so many suicides, I don't think any of us who loved him could have prevented him from doing what he did.

My mother chose not to have a memorial service for my dad. She was probably worried about what their friends and neighbors would think, and was ashamed of what he had done. I tried to gently talk her out of her decision, but I recognized that it was hers to make.

And so Lorrie and I, my sister and her husband, along with my mom and a young minister, gathered after his death to scatter his ashes across our property in front of Lake Texoma.

It was a cold, bleak, gray day. In Texas, in the winter, the grass is dormant and brown. It all felt so lonely.

I said a few words. My sister said something. So did the minister, who had driven up from Waples Memorial United Methodist Church in Denison. When it was my mom's turn, her words were simple: "I had a chance to say everything I needed to say to him when he was alive. There was nothing left unsaid." My mother was outwardly OK, strong and stoic.

None of us spoke too long. I guess we were just shocked standing there, and angry that my father had made that choice. I was especially upset that he would choose to remove himself from my daughters' lives. I couldn't believe he would do that.

After Flight 1549, people wrote to tell me that they could sense how much I valued life. Quite frankly, one of the reasons I think I've placed such a high value on life is that my father took his.

I didn't think about my father's suicide when I was in the cockpit of Flight 1549. He wasn't anywhere in my thoughts. But his death did have an effect on how I've lived, and on how I view the world. It made me more committed to preserving life. I exercise more care in my professional responsibilities. I am willing to work very hard to protect people's lives, to be a good Samaritan, and to not be a bystander, in part because I couldn't save my father.

After my father died, and my mom was able to come to terms with her grief and guilt, she reinvented herself. I was very proud of her. She traveled, and after a few years, she even met a nice man and began dating him seriously. She really blossomed.

I think my mother would have continued to live a rich and busy life if she hadn't been diagnosed with colon cancer in December 1998.

The day I got the news of her cancer, I was finishing a trip on the MD-80 in Pittsburgh, and I immediately got on a flight to Dallas. My mother knew she was terminal, and said so. It was shocking for us. She was only 71 years old and had never been seriously ill in her life. She came from a line of long-lived people. Her father lived until age 94 and her mother until 102.

But we accepted the hand she'd been dealt, and in my mother's final weeks, I had a chance to have many talks with her about our lives, about her wishes for Kate and Kelly. She said she had few regrets. Unlike with my dad, I was able to say good-bye. My mom lived just one month after her diagnosis. And so for the second time in just a few years, we experienced a heartbreaking loss. This time, I felt all the things I had felt after my father's death, except anger.

There have been lessons for me.

In the three years between my father's suicide and my mom's death, my mother was severely tested. But the former schoolteacher taught herself how to get the most out of life and how to be as happy as possible. I admired her even more for how she lived as a widow.

I didn't think of her when I was in the cockpit of Flight 1549, but her will to live had already served as an inspiration to me.

Lorrie and I wish my parents could have lived to witness what has happened as a result of Flight 1549. The incident would have been frightening for my mother, and very emotional. She'd be overjoyed at the outcome, of course. My mother would have cried. My father would have been proud.

When I first became a pilot, my mother was always telling me to stay safe. "Fly low and slow," she'd say. I'd roll my eyes. It was like a comedy routine between us.

I'd remind her that flying low and slow isn't as safe as flying higher and at an appropriate speed. She understood that. But the line "fly low and slow" became her way of encouraging me to be careful. It was her handy little admonition.

We were certainly flying low over the Hudson on January 15. Without engines, we were slowing down, too. I can imagine my mom would have had a comment of some kind: "Low and slow turned out OK for you, didn't it?"

I assume my father would have summed up Flight 1549 by telling me something like: "It looks like you learned your lessons well. You became good at something you cared about it and it paid off. You made a difference."

I don't know if he would have bought into any of the hero accolades thrown my way. In his generation, people were put in tough situations and they were up to the task. His contemporaries won World War II, and for the most part, did it humbly and without personal aggrandizement. I think my dad would have been proud of my achievements, but he would have put what happened in perspective: I did my job well. So have a lot of other people before me.

My father and I were affectionate, and we were close in our own somewhat stiff way. But we weren't as close as I wish we could have been. That was his temperament and mine. We were both quiet and pretty stoic. We never shared a lot of personal feelings. We kept a lot to ourselves.

There wasn't really any yelling and screaming in our house; we were all too polite and reticent. That made for a calm childhood, but there was a flip side to that. Though we enjoyed each other's company, we didn't share a great deal of emotion. We didn't talk about too many personal things. As I got older, a part of me envied and admired those big, stereotypical ethnic families where people argued all the time, almost as a way of showing love. I didn't grow up in a family where everyone was always offended and making grand, dramatic pronouncements. Don't get me wrong. It was wonderful to be in a peaceful household. But it could also feel slightly passionless at times.

I think that the urges toward staid family dynamics are in my DNA. I've tried to broaden myself and break out of the mold with my daughters, to be more outwardly emotional. I'm still working on it.

Kate and Kelly were toddlers when my parents died, and I wish my mom and dad were alive to see the lovely young women they have become. I have tried to pass on my parents' values to them, and I can see that the girls have embraced many of them.

The girls also have attributes and gifts that come from within them. It's not that Lorrie and I have taught them, or that we've even shown them the way. And in the wake of Flight 1549, some of these attributes of theirs have become clearer to me.

Kate, for instance, is supremely self-confident. When Lorrie and I reflect on how comfortable Kate is with herself, we sometimes say we want to grow up to be just like her. Now sixteen years old, she is also very focused and funny, and she is a conscientious student. She has always wanted to be a veterinarian and has never wavered.

Her friends say she may be the most self-assured kid they know. They have stories about her that prove their point. Once, in middle school, a girl didn't like the shirt Kate was wearing and told her so. "I'm sorry you don't like it," Kate answered, "but I like it a lot."

Lorrie says many girls would have dissolved in the wake of a peer's dismissive fashion comment. Not Kate.

She's comfortable around boys, too. Once, when she was nine years old, we were on vacation at a ski resort and she saw a bunch of older boys making a snowman. "I'm going to go play with them," she told us.

We cautioned her. She didn't know any of them. They were a few years older. But she marched fearlessly right into that circle of boys and announced she was there to play. She staked her claim. At first the boys looked shocked. And then, because she was so sure of herself, they let her join them for the rest of the afternoon. Lorrie and I marveled at her confidence.

A few weeks after Flight 1549, I saw that confidence again, when she took her driver's license test at the California Department of Motor Vehicles. Lorrie and I went along, and we were both nervous for her. She had prepared well, and I trusted her behind the wheel, but you never know how a kid will perform in the tension of the moment.

While Kate took her road test, Lorrie and I stayed behind in the DMV waiting area. It felt like a long twenty-five minutes before she returned with a big smile on her face. She had passed.

I had to ask her: "Was it hard? Were you worried you'd fail?"

Her answer: "I knew I could do it."

What Kate meant was this: She was confident because she had done all the preparation. She had worked and studied and practiced.

When she said that, she reminded me of how I felt when the engines died on Flight 1549. In fact, she had used the exact same words I had used when Katie Couric asked me whether I was confident while descending toward the Hudson. Kate didn't remember those were my words on TV. She just had the same confidence in her preparation.

Kate has always seen things in black and white. It's yes or no. It is or it isn't. Lorrie says she's like me in that way. She has always been very controlled with her emotions, very much the intellectual. I understand that about her, and even though we're alike, it's not always easy for us to connect emotionally.

For a couple of years now, Kate's growing independence has been tough for me. As she became a teenager, she was less willing to confide in me. She'd still turn to Lorrie, but I sometimes felt like an outsider. Her old dad.

Flight 1549 changed the dynamics a little. She's willing to be more physically affectionate now. The love between us often remains unspoken, but we both feel the connection intensely.

Unlike Kate, fourteen-year-old Kelly has always been very sensitive and affectionate. As a toddler, Kelly would snuggle up with us—Lorrie called her "our snuggle bunny"—and it was just the greatest feeling. She also would be more apt to cry when I left on a trip. When she was three or four years old, and she'd see me putting on my uniform, the tears would well up.

Kelly has always been innately empathetic. If there's a new girl at school or a child with disabilities, she is the first one to arrange a playdate or to say, "Why don't you sit with us at lunch?" She always feels a need to reach out to these kids, and it can be an emotional burden for her.

Given how deeply she feels things, she is sensitive to words that sting. She doesn't engage in the some-times rough dialogue that is normal for teenagers. She takes greater care with her words. She will couch even something negative in gentler terms. She doesn't want to hurt people's feelings.

I remember when she would get home from school in third or fourth grade, and Lorrie and I would ask her, "So how was your day?"

Invariably, she'd tell us about a schoolmate who was having a tough day at school. She could sense when someone else was troubled. She felt this need to reach out to them. I know that can be an emotional burden for her.

From day one after Flight 1549, Kelly experienced the incident fully. The moment Lorrie told her what had happened, she started to cry, even though she al-ready knew I was safe. Her feelings were partly rooted in the idea that my life had been at risk. But I also think she deeply felt what that experience must have been like for me, and her heart went out to me. Hearing the details was very disturbing to her.

Both Kelly and Kate saw their grades take a hit in the wake of Flight 1549, and Kate wasn't able to get hers back up completely. At first, it was a stressful time for all of us. They missed school and then, as soon as they returned, took several exams that they weren't prepared for. Once they were in that deep hole, it got hard to get their averages back up. Our routine was disrupted for weeks, and the "public figure" aspect of our new lives—always having to be "on" when we were in public—was hard for them.

In the wake of the flight, we've sat down together as a family to read through some of the stacks of mail we've received from around the world. It helped us process the event together, to see how other people connected with it emotionally. It reminded us to cherish the bonds between us, because nothing is ever for sure. I think the girls have a better understanding of this now.

As teenagers, Kate and Kelly are far less apt to snuggle with Lorrie and me than they once were. We miss that. Sometimes, when they're not feeling well, it becomes OK to snuggle again. And in the wake of Flight 1549, we hug a bit more. I'm more apt to kiss the girls before I leave town, even if it is early in the morning and they're in bed, sleeping.

A few weeks after Flight 1549, Lorrie wrote a letter of thanks to all the friends and strangers who had gotten in touch with her to express their concern. "It is still hard for me to sort out all my emotions," she wrote. "The events of January 15 have been like an onion, multilayered, and peeling back the layers has taken time and will take more time to come. For me, there was the accident itself, the huge media interest, and then the mail.

"It's interesting how our brains protect us from trauma, because after Sully told me the news, I didn't feel panicked. I just felt this weird, out-of-body feeling that it was not real. I was going through the motions but I could not believe that the images I was seeing on TV were of my husband's plane.

"I know intellectually and believe with all my heart that commercial aviation is the safest form of travel, so I have never been afraid of Sully's career. How incredible were the odds that my husband was involved in an airline accident? Impossible, and yet not."

Flight 1549 has had an impact on our marriage. The resulting emotions for both of us have been overwhelming and sometimes confusing, and we haven't been able to sufficiently be there for each other at every step.

One morning, five months after the incident, Lorrie said to me, "I've wanted to cry all morning." And so she went by herself to our favorite hill in the neighborhood—the "anything is possible" hill. She stood on top, took a moment that was all her own, and cried. Why was she crying?

"The accident, the aftermath, it's still unbelievable to me," she told me. "I feel like I haven't been able to fully process it all."

It isn't just that Flight 1549 jolted her into the realization that she could lose me. "I've always known I could lose you," she says. "Like all of us, you're at the mercy of those driving next to you on the highway, or the food you're eating in a restaurant, or a disease we don't yet know about. So it's not that I feel like you're cheating death every time you fly."

Instead, Lorrie just feels as if the incident in the Hudson, and the continuing aftermath, has scrambled her brain. It affected the dynamics in our family.

For our entire marriage, Lorrie spent long stretches as a single parent. I'd be off on trips, and she'd be dealing with everything in the household. It seemed like things always decided to break when I was gone— the car, the washing machine, the oven. Once, I was on a flight doing preparations before pushing back from the gate, and my cell phone rang. It was Lorrie in a panic. Water was pouring down the side window of our house. At first she thought it was a bad storm, but then she realized that the seal on our pool pump had broken, and water was gushing into the air like an open fire hydrant.

"Oh my God!" Lorrie said. "The pool is broken! A quarter of the water that was in it has drained out already, and hundreds of gallons are raining down on our window!"

"I'm about to push back," I said to her, which meant I was required to turn off my cell phone. "Turn off the filter pump and call the pool guy. I have to go. I'm sorry." And then I shut off my cell phone, taxied toward the runway, and left her on her own to stop the rain.

No woman dealing with an emergency like that wants her husband hanging up on her. Again and again, my flying career came at a cost.

I've been even busier and more out-of-pocket since Flight 1549. I've been asked to make appearances, give testimony, answer requests from the media, and travel as a public face of the piloting profession. For the first seven months after the Hudson incident, I wasn't even flying planes for US Airways. Still, some weeks, I'd be gone from home more than I used to be when I was in the cockpit.

"You won't get a do-over with the girls," Lorrie has been telling me. "If you wait until the next year or the year after that to live your family life, you'll miss too much. The time you've lost is gone forever."

I know this, and I've tried to make adjustments in my life.

A stressful incident such as Flight 1549 either pulls a couple closer together or leaves them further apart. Lorrie and I have seen both extremes. At first, we clung to each other like ports in a storm. There was an onslaught of attention, and we were hanging on to each other for dear life.

Now Lorrie sometimes gets frustrated with me when I'm "Sully, the public figure." Almost everywhere I go, people recognize me and want to interact, get an autograph, or reflect on something from their own lives. I'm cordial and gracious to everyone, and genuinely interested in their stories. Sometimes, when I get home, I can be frazzled and used up and short-tempered. I can be impatient with the girls.

"You have your priorities wrong, Sully," Lorrie has told me firmly. "As nice as you are to strangers, that's the same nice you need to be to me and the girls."

She is completely right about that, and I'm lucky to have a spouse who loves me enough to tell it to me straight.

At about eight o'clock one morning, a few months after Flight 1549, Lorrie and I were in our garage, looking out into the street. Kate had just pulled out of the driveway, headed for school. It was a bright, beautiful morning, but inside the garage, we were standing in shadow. Lorrie and I were holding hands and watching her pull away.

Kate began her three-point turn to pull out of our court, and she stopped for a moment to shift from reverse into drive. As she turned her head, her ponytail was swaying, and she looked so grown up. She looked almost like a woman in her twenties. It was startling to us.

In that instant, I felt a cascade of images coming into my head, images of her growing up and becoming the strong, confident young lady she now is. It was almost as if she were driving away that morning on her way to her own adult life. Standing there, I remembered when we took her to her first day of preschool at St. Timothy's Episcopal Church in Danville, and how a lot of the other kids were clinging and crying, and Kate just took off, happily independent. She said good-bye and never looked back.

In that moment, I also thought about an essay Kelly wrote in third grade. In the spring of 2002, US Airways had parked its MD-80 fleet and was retraining pilots on the Airbus. Until I got the Airbus training, I wasn't flying, and I was able to remain home for a few months, very present in the kids' lives. Kelly's essay assignment, in the fall of 2002, was to write about the happiest time of her life. "The happiest time of my life," she wrote, "was the time when Daddy was home." Reading that was one of those bittersweet moments that filled and broke my heart at the same time.

Now here we are, with the girls pulling out of our driveway all on their own. I've blinked and everything has changed: My parents are long gone, the things I missed with my kids can't be reclaimed, and my life is different now. Lorrie is right. I need to remember every day how precious our time with the girls really is.

By landing safely, Flight 1549 returned passengers and crew to the loving embrace of their families. We've all been given second chances. We've been given new reminders that we are loved, and new opportunities to show affection to those we care about. There were 155 people on that plane who got to go home. I must never lose sight of the fact that I was one of them.

19

THE QUESTION

One day in early May, almost four months after Flight 1549 landed in the Hudson, three large cardboard boxes arrived at my front door in Danville. Inside, well preserved and neatly packaged, were the things I had left behind in the cockpit of the plane. Everything was there except that eight-dollar tuna sandwich I had bought and never eaten before takeoff.

I was somewhat solemn going through my belongings. I knew that after most airline accidents, such boxes are sent to relatives of victims who've died. Or else, when a plane crashes, fire destroys most everything, or the victims' belongings have been shattered into pieces so small that there is almost nothing to be returned. Maybe relatives will get back someone's wedding ring. Usually loved ones get little or nothing.

In the case of Flight 1549, all of us who were "survivors" got boxes addressed directly to us. We were able to sign the FedEx slips ourselves. Some of what was returned to us was destroyed and unusable. But a lot of things were in good condition and could be folded back into our lives. Passengers got back their favorite jeans, their coats, their car keys, their purses. I pictured these passengers, all over the country, opening their boxes and flashing back to January 15, 2009. We could focus on waterlogged items that were ruined, or we could go through our personal effects feeling grateful.

The plane had sunk into the Hudson after we all evacuated, and a company from El Segundo, California, Douglass Personal Effects Administrators, was charged with taking what was fished out of the water and trying to reclaim what they could. I was impressed by the job they undertook in order to reunite us with our belongings. They went through every suitcase in the cargo hold and every item in the overhead compartments.

It was amazing and impressive that so many things submerged in dirty, icy water could be brought back to life. The company used sheets of fabric softener to separate all of the clothing and other items. The smell of dryer sheets was overpowering when we opened our boxes.

My roll-aboard bag was in one of the boxes, its contents dried, inventoried, and wrapped up in tissue paper. My iPod, laptop, and alarm clock were trashed. But my phone charger and iPod charger still worked. So did my data cable for transferring photos from my phone to my computer. My mini Maglite also worked fine. My running shoes looked as good as new. The shoes I was wearing on the flight came home with me in January but were totally waterlogged and beaten up. I really hoped they could be saved, because they were what we call "airport-friendly shoes," with no metal; I didn't have to take them off to go through security checkpoints. I took those shoes to my favorite local shoe repairman at a shopping center in Danville, and he did a wonderful job fixing and cleaning them up. I wear them still.

On January 15, I was traveling with four library books, including a copy of *Just Culture*, a book about safety issues. I later called my local library to apologize for leaving the books on the plane, and they agreed not to charge me for replacing them.

Anyway, I was glad to find all four of the library books in one of the boxes of my belongings. The reclamation company had tried using a drying process to make the books usable again but weren't completely successful. The pages are readable but too wrinkled to be checked out again by library patrons. I returned them anyway. The library has found a place for them to be displayed.

Since Flight 1549 came at the end of a four-day trip, I had mostly dirty laundry in my roll-aboard bag. All of my clothing came back in good condition, ready to wear, and with that strong fabric-softener smell.

I was also glad to get back my Jeppesen airway manual, which contains the charts for all of the airports we serve. Still taped neatly inside the manual, weathered but readable, was the fortune from a fortune cookie that I'd gotten at a Chinese restaurant in San Mateo, California, sometime in the late 1980s.

The fortune read: "A delay is better than a disaster."

I thought that was good advice at the time and so I'd kept it in the manual ever since.

That fortune reminded me of an unexpected question Kate asked me when she was nine years old. I was driving her to school, and out of the blue, she asked me: "Daddy, what does *integrity* mean?"

After thinking about it for a little bit, I came up with what, in retrospect, was a pretty good answer. I said, "*Integrity* means doing the right thing even when it's not convenient."

Integrity is the core of my profession. An airline pilot has to do the right thing every time, even if that means delaying or canceling a flight to address a maintenance or other issue, even if it means inconveniencing 183 people who want to get home, including the pilot. By delaying a flight, I am ensuring that they will get home.

I am trained to be intolerant of anything less than the highest standards of my profession. I believe air travel is as safe as it is because tens of thousands of my fellow airline and aviation workers feel a shared sense of duty to make safety a reality every day. I call it a daily devotion to duty. It's serving a cause greater than ourselves.

And so I think often of that fortune, which sat for a good while in the cockpit of a water-filled Airbus A320, tilted sideways in the Hudson: "A delay is better than a disaster."

It's nice to have that fortune back. It will definitely accompany me on future flights.

A few days after receiving my belongings, I flew to Washington, D.C., where I met Jeff Skiles at the headquarters of the National Transportation Safety Board. We had been invited to listen to the cockpit voice recorder (CVR), and to offer our thoughts and memories.

Previously, the only tape available had been from the FAA, and that contained the radio communications between us and Air Traffic Control. This NTSB visit would be our first opportunity to listen to the audio from the cockpit voice recorder. We'd hear exactly what we had said to each other in the cockpit during the flight. For four months until this May meeting, both of us had been relying on our memories of what we had said. Now, finally, we would know for sure.

There were six of us in the room: Jeff Skiles, Jeff Diercksmeier, a U.S. Airline Pilots Association accident investigation committee member, three NTSB officials (two investigators and a specialist from the agency's recordings section), and me. The investigators were happy to have Jeff and me there with them. After many airline accidents, when the recordings are reviewed, the flight crews are not on hand. Often, the pilots whose voices are on the recordings are dead, and so they can't explain what they were thinking, why they made the decisions they did, or exactly what a particular word was.

Listening to the tape was an intense experience for us. It brought us back together into the cockpit, as if we were reliving the incident in real time.

We were in a small office with fluorescent lights, and we sat in chairs at a table, wearing headsets. Jeff and I didn't look at each other much. For the most part, we were in our own heads, often with our eyes closed, trying to capture all the sounds and noises in the cockpit.

The recording began while Flight 1549 was about to push back from the gate and continued until we first touched the Hudson. There were things I said on the tape that I didn't recall saying. Just thirty-three seconds before the bird strike, I said to Jeff, "And what a view of the Hudson today!" He took a look and agreed: "Yeah!"

The bird strikes were completely audible on the tape. There were the sounds of thumps and then unnatural noises as the birds went through the engines. You could hear the damage being inflicted on the engines, and how they protested with sickening sounds that an engine should never make. We clearly heard the *wooooooh* of engines spooling down and rolling back, followed by the sounds of vibrations as the engines tore themselves apart. Listening to the tape, I was reminded of how we felt in that moment. It was as if the bottom were falling out of our world. Even in the safety of that office at the NTSB, it was disturbing for us to hear again the rundown of the engines, and to know we had been in the cockpit of that aircraft when that was occurring.

The biggest surprise for me, listening to the tape, was how fast everything happened. The entire flight was five minutes and eight seconds long. The first minute and forty seconds were uneventful. Then, from the moment I said, "Birds!" until we approached the water and I said, "We're gonna brace!" just three minutes and twenty-eight seconds had passed. That's less time than it takes me to brush my teeth and shave.

The whole incident took a bit longer in my memory. Yes, I knew and felt all along that things happened fast. But in my recollections, it was as if I had a little more time to think, to decide, to act—even if it was abbreviated.

Listening to the tape, however, I realized that everything really happened in 208 extraordinarily time-compressed seconds. Frankly, it was beyond belief. Beyond extreme. It was overwhelming. It took me right back to the moment. I didn't tear up, but I know there were muscle changes in my face as I listened. It was surprising and emotional for Jeff, too.

Somehow, time must have slowed down in my head that day. It's not as if everything was in slow motion. It's just that, in my memory, it didn't feel as incredibly fast as the tape made obvious that it was.

There are different microphones in the cockpit, which can pick up voices, noises, warning chimes, and radio transmissions, including those from other planes. The NTSB was able to play back whatever was picked up by each microphone, one at a time, so we could isolate certain sounds and hear things that were at first masked by louder sounds. The investigators asked us to explain sounds or snippets of conversation that weren't clear on the tape.

I was very happy with how Jeff and I sounded on the tape, and how we handled ourselves individually and as a team. We did not sound confused and overwhelmed. We sounded busy. I've read many transcripts of accidents over the last thirty years, and this one sounded really good in terms of our competence.

Jeff and I had met just three days before we flew Flight 1549. Yet during this dire emergency—with no time to verbalize every action and discuss our situation—we communicated extraordinarily well. Thanks to our training, and our immediate observations in the moment of crisis, each of us understood the situation, knew what needed to be done, and had already begun doing our parts in an urgent yet cooperative fashion.

Departure control (3:28:31): *"All right, Cactus fifteen forty-nine it's gonna be left traffic for runway three one."*

Sullenberger on radio (3:28:35): *"Unable."*

Traffic Collision Avoidance System in cockpit—synthetic voice oral warning (3:28:36): *"Traffic! Traffic!"*

Departure control (3:28:36): *"Okay, what do you
need to land?"*

Predictive Windshear System synthetic voice
(3:28:45): *"Go around. Wind shear ahead."*

Skiles (3:28:45): *"FAC-1 [Flight Augmentation
Computer 1] off, then on."*

Skiles (3:29:00): *"No relight after thirty seconds,
engine master one and two confirm off."*

Sullenberger (3:29:11): *"This is the captain. Brace for
impact!"*

Forty-four more seconds passed, with Jeff and me
engaged in challenge-and-response as we went through
the checklist while listening to both Patrick the controller and the repetitive chimes of the flight warning
computer.

Enhanced Ground Proximity Warning System
synthetic voice (3:29:55): *"Pull up. Pull up. Pull up.
Pull up. Pull up. Pull up."*

Skiles (3:30:01): *"Got flaps out!"*

Skiles (3:30:03): *"Two hundred fifty feet in the air."*

As I listened to the recording, I saw clearly that Jeff
was doing exactly the right things at exactly the right
moments. He knew intuitively that because of our short

time remaining before landing and our proximity to the surface, he needed to shift his priorities. Without me asking, he began to call out to me the altitude above the surface and the airspeed.

> Enhanced Ground Proximity Warning System synthetic voice (3:30:24): *"Terrain terrain. Pull up. Pull up. Pull up. Pull up. Pull up. Pull up . . ."*
> Sullenberger (3:30:38): *"We're gonna brace!"*

It was awful and beautiful at the same time.

Jeff and I had found ourselves in a crucible, a cacophony of automated warnings, synthetic voices, repetitive chimes, radio calls, traffic alerts, and ground proximity warnings. Through it all, we had to maintain control of the airplane, analyze the situation, take step-by-step action, and make critical decisions without being distracted or panicking. It sounded as if our world was ending, and yet our crew coordination was beautiful. I was very proud of what we were able to accomplish.

After Jeff and I heard the recording for the first time with the NTSB investigators, we excused ourselves to go to the men's room. We would have to listen to the tape several more times on this day, but I think we both wanted a break before we did that.

As we walked down the hallway of this old government office building, I turned to Jeff and asked, "What did you think?"

Before he could answer, I felt a need to say something. "I'll tell you what I think," I told him. "I'm so proud of you. Within seconds of me calling for the checklist, you had it out, you found the right page, you had begun reading it. And you were right there with me, step-by-step, challenge-and-response, through all of those distractions. We did this together."

In the media, I'd gotten most of the credit for Flight 1549. "I don't care what anybody says," I told Jeff. "We were a team."

He looked at me, and I saw tears in his eyes. "Thank you," he said. I was a bit choked up myself. We hugged, then stood together for a moment in that hallway, not saying anything. We were two men who'd been through something extraordinary together and couldn't find the words to fully capture it.

Eventually, we made our way back to the CVR lab, where we joined the investigators and listened to the cockpit recording again and again.

When Kelly was very young, she once asked me, "What's the best job in the world?"

My answer to her was this: "It's the job you would do even if you didn't have to." It's so important for people to find jobs suited to their strengths and their passions. People who love their jobs work more diligently at them. They become more adept at the intricacies of their duties. They serve the world well.

On January 14, 2009, my life had been a series of thoughtful opportunities to be the best pilot, leader, and teammate I could be. I was an anonymous, regular guy—a husband, a father, a US Airways pilot. On January 15, circumstances changed everything, a reminder that none of us ever knows what tomorrow will bring.

I flew thousands of flights in the last forty-two years, but my entire career is now being judged by how I performed on one of them. This has been a reminder to me: We need to try to do the right thing every time, to perform at our best, because we never know which moment in our lives we'll be judged on.

I've told Kate and Kelly that each of us has the responsibility to prepare ourselves well. I want them to invest in themselves, to never stop learning, either professionally or personally. At the end of their lives, like all of us, I expect they might ask themselves a simple question: Did I make a difference? My wish for them is that the answer to that question will be yes.

As for myself, I look back at everything and continue to feel lucky. I found my passion very early. At five years old, I knew I would spend my life flying. At sixteen, I was already in the sky alone, practicing and practicing, circling happily above Mr. Cook's grass strip.

In the years that followed, my romance with flying helped sustain me. At twenty-four, I was a fighter pilot, learning that I had to pay the closest attention to everything, because life and death could be separated by seconds and by feet. By fifty-seven, I was a gray-haired man with my hands on the controls of an Airbus A320 over Manhattan, using a lifetime of knowledge to find a way to safety.

Through it all, my love of flying has never wavered. I'm still that eleven-year-old boy with his face pressed against the window of the Convair 440, ready to take my first ride out of Dallas on an airplane. I'm still that earnest teen who flew low over our house on Hanna Drive, waving to my mom and sister on the ground. I'm still the serious young Air Force cadet, in awe of all the fighter pilots who came before me and showed me the way.

Just as I completely love Lorrie, Kate, and Kelly, I will never shake my love of flying. Never.

At the moment, I'm not sure exactly what my next steps in life might be. Where will flying take me next? What tests are ahead? What opportunities? I do know that I will continue to be an airline pilot. It's part of what gives me purpose. It's a big part of who I am.

I'm sure there will be passengers on future US Airways flights who will look toward the closed cockpit doors and wonder: Who is flying this plane today? Most likely, the captain will be one of my colleagues, an aviator who is well disciplined and well trained, with the highest sense of duty and a great love of flight.

Then again, the guy behind that door may be me. Once we're in the air, I'll say a few words about the cruising altitude, the flying time, and the weather. I'll remind passengers to keep their seat belts fastened, because turbulence often comes unexpectedly. And then I'll switch off the public address system, and I'll do my job.

Acknowledgments

I could not have written this book without the support of my family. Kelly, Kate, and Lorrie have always been there for me with their thoughtfulness, love, and kindness. I know that every moment I spent writing was a moment I could not spend with them, which made this project all the more difficult. I am grateful for your understanding in granting me the time I needed to write this book.

The best preparation for this event was to have the right partner in my life. I wish everyone could find someone as smart, caring, supportive, independent, well-spoken, and strong as Lorraine Sullenberger. Lorrie, I couldn't have made it through the aftermath of January 15 without you at my side and in my heart.

My mother and father taught me about hard work, integrity, and lifelong education. I am grateful to them for instilling in me a set of values which have been constant guideposts throughout my life. I also thank my sister, Mary, for her love and support.

On January 15, 2009, First Officer Jeff Skiles and I found ourselves in a crucible where we were fighting for our lives and the lives of all our passengers and crew. We worked together closely from start to finish, and our effective teamwork was essential in achieving a successful outcome. Jeff, you have my eternal gratitude for your skill and bravery.

Jeff and I were joined on Flight 1549 by flight attendants Donna Dent, Doreen Welsh, and Sheila Dail, whose instinctive and immediate collaboration in a time of crisis kept the passengers calm and helped us overcome the challenges we faced. I continue to be impressed with your strength and steadfastness since that day.

I thank the people of Denison, Texas, who helped shape me as a youth, and the people of Danville, California, whom I am proud to count as neighbors and friends. I also want to thank the people of New York and New Jersey, especially NY Waterway, the New York Police Department, the United States Coast Guard, the Fire Department of New York, the Federal

Bureau of Investigation, the Port Authority of New York and New Jersey, and the New York City Office of Emergency Management. I owe a debt of gratitude to all those who played a role in saving our lives on January 15.

Thank you to Lorrie's friends Tamara Wheeler, Margaret Combs, Bunny Martin, Kathy Giger, and Heather Hildebrand. In the hours following the Hudson landing, when I was attending to my duties in New York and could not be with my wife and girls, these women helped my family through the sudden and overwhelming media attention.

While I've read my fair share of books over the years, I never thought that I'd find myself writing one, and Jeff Zaslow has been a remarkable partner throughout this endeavor. I am thankful for his assistance, his investigative skills, his instincts as a veteran reporter, and his unfailingly sage advice.

The team at HarperCollins did a great job of guiding this first-time writer through the process. I'd like to thank David Highfill, Seale Ballinger, Sharyn Rosenblum, and the entire HarperCollins team that helped me get this project off the ground and onto the bookshelf.

My literary agent, Jan Miller, and her associate Shannon Marven have also offered tremendous advice and counsel. They and their colleagues at Dupree/ Miller helped me find my way to HarperCollins and deftly guided me through the process of taking a book from idea to completion.

Since the day after the event, Alex Clemens, Libby Smiley, and their colleagues at Barbary Coast Consulting have been by my family's side, guiding us through this unfamiliar territory with their wise counsel and tireless efforts.

Thanks also to Gary Morris, Captain James Hayhurst, Alex King, Captain Al Haynes, Helen Ott, Bracha Nechama Bomze, Herman Bomze, Patrick Harten, Eric Stevenson, Conrad Mueller, Paul Kellen, Karen Kaiser Clark, Bart Simon, Theresa Hunsicker, and David Sontag.

My union colleagues were an incredible source of support on January 15, 2009, and throughout the aftermath. Thank you especially to Captain Larry Rooney and Captain Dan Sicchio, who have spent countless hours assisting me with everything from my NTSB testimony to this book. Thanks also to First Officer Gary Bauhan, Captain Ken Blitchington, Captain Steve Bradford, Captain Dan Britt, Captain John Carey, Captain Carl Clarke, Captain Mike Cleary, First Officer Jeff Diercksmeier, Captain Peter Dolf, Captain David Douglas, Captain Arnie Gentile, First Officer

Bob Georges, Captain Michael Greenlee, Captain Pete Griffith, Captain Jonathan Hobbs, Captain Mark King, Captain Tim Kirby, Captain Tom Kubik, Dr. Pete Lambrou, Captain Jan Randle, Captain James Ray, Captain John Sabel, Lee Seham, First Officer Carol Stone, Captain Gary Van Hartogh, Captain Valerie Wells, and Captain Lucy Young. Each of you was there for me at a time when I very much needed your help. I am indebted to you, and to all my brothers and sisters in the U.S. Airline Pilots Association.

I'd like to thank all the people who work at US Airways. You have consistently confronted the challenges facing our profession with grace and excellence, and I am proud to call you my colleagues. All airline employees have an important job to do, and despite changes in the industry, they do it well. Readers, I hope that the next time you fly, you take a moment to thank your flight attendants for continually preparing for your safety, and your pilots for the dedication and care with which they conduct each and every flight.

Thank you to my U.S. Air Force Academy classmate and retired Northwest Airlines captain Mike Hay and my fellow fighter pilot and current Southwest Airlines captain Jim Leslie for their assistance in reviewing the events in this book and supplementing my memories with their own. While their help has been invaluable throughout the writing process, I take responsibility for the content of this book. Any errors or omissions are mine alone.

And finally, I'd like to thank L. T. Cook Jr., who saw the potential in me and helped me realize it.

Flight Path of Flight 1549, January 15, 2009

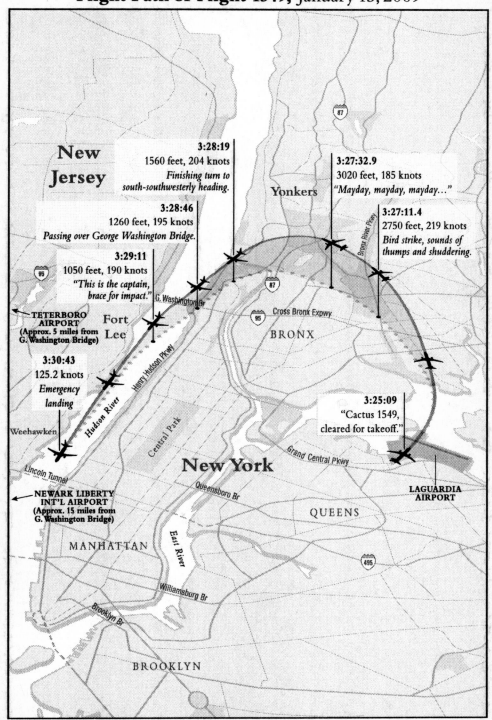

New
Jersey

3:28:19
1560 feet, 204 knots
*Finishing turn to
south-southwesterly heading.*

3:27:32.9
3020 feet, 185 knots
"Mayday, mayday, mayday…"

Yonkers

3:28:46
1260 feet, 195 knots
Passing over George Washington Bridge.

3:27:11.4
2750 feet, 219 knots
*Bird strike, sounds of
thumps and shuddering.*

3:29:11
1050 feet, 190 knots
*"This is the captain,
brace for impact."*

G. Washington Br

Cross Bronx Expwy

BRONX

**TETERBORO
AIRPORT**
**(Approx. 5 miles from
G. Washington Bridge)**

Fort
Lee

3:30:43
125.2 knots
*Emergency
landing*

Henry Hudson Pkwy

Hudson River

3:25:09
*"Cactus 1549,
cleared for takeoff."*

Weehawken

Central Park

Grand Central Pkwy

New York

**LAGUARDIA
AIRPORT**

Lincoln Tunnel

**NEWARK LIBERTY
INT'L AIRPORT**
**(Approx. 15 miles from
G. Washington Bridge)**

Queensboro Br

QUEENS

MANHATTAN

East River

Williamsburg Br

Brooklyn Br

BROOKLYN

Appendix B

Transcript of an Allied Signal/Honeywell model SSCVR cockpit voice recorder (CVR), s/n 2878, installed on an Airbus Industrie A320-214, registration N106US. The airplane was operated by US Airways as Flight 1549, when it ditched into the Hudson River, NY, on January 15, 2009.

LEGEND

RDO   Radio transmission from accident aircraft, US Airways 1549

CAM   Cockpit area microphone voice or sound source

PA   Voice or sound heard on the public address system channel

HOT	Hot microphone voice or sound source[1]
TOGA	Takeoff/Go Around thrust
INTR	Interphone communication to or from ground crew

For RDO, CAM, PA, HOT and INTR comments:

-1   Voice identified as the Captain (Hot-1: Capt. Sullenberger)

-2   Voice identified as the First Officer (Hot-2: First Officer Skiles)

-3   Voice identified as cabin crewmember

-4   Voice identified as groundcrew

-?   Voice unidentified

FWC	Automated callout or sound from the Flight Warning Computer
TCAS	Automated callout or sound from the Traffic Collision Avoidance System
PWS	Automated callout or sound from the Predictive Windshear System
GPWS	Automated callout or sound from the Ground Proximity Warning System

---

1. This recording contained audio from Hot microphones used by the flight-crew. The voices or sounds on these channels were also, at times, heard by the CVR group on the CAM channel and vice versa. In these cases, comments are generally annotated as coming from the source (either HOT or CAM) from which the comment was easiest to hear and discern.

EGPWS	Automated callout or sound from the Enhanced Ground Proximity Warning System
TWR	Radio transmission from the Air Traffic Control Tower at LaGuardia
DEP	Radio transmission from LaGuardia departure control (Air Traffic Control Specialist Harten)
CH[1234]	CVR Channel identifier 1 = Captain 2 = First Officer 3 = PA 4 = Cockpit Area Microphone
★	Unintelligible word
@	Non-Pertinent word
&	Third party personal name (see Note 5, p. 398)
#	Expletive
-, —	Break in continuity or interruption in comment
( )	Questionable insertion
[ ]	Editorial insertion
. . .	Pause

Note 1: Times are expressed in Eastern Standard Time (EST), based on the clock used to timestamp the recorded radar data from the Newark ASR-9.

Note 2: Generally, only radio transmissions to and from the accident aircraft were transcribed.

Note 3:   Words shown with excess vowels, letters, or drawn out syllables are a phonetic representation of the words as spoken.

Note 4:   A non-pertinent word, where noted, refers to a word not directly related to the operation, control or condition of the aircraft.

Note 5:   Personal names of 3rd parties not involved in the conversation are generally not transcribed.

Intra-Cockpit Communication		Air-Ground Communication	
Time and Source	**Content**	**Time and Source**	**Content**
		15:24:54 TWR	Cactus fifteen forty nine runway four clear for takeoff.
		15:24:56.7 RDO-1	Cactus fifteen forty nine clear for takeoff.
15:25:06 CAM	[sound similar to increase in engine noise/speed]		
15:25:09 CAM-2	TOGA.		
15:25:10 HOT-1	TOGA set.		
15:25:20 HOT-1	eighty.		
15:25:21 HOT-2	checked.		
15:25:33 HOT-1	V one, rotate.		
15:25:38 HOT-1	positive rate.		
15:25:39 HOT-2	gear up please.		
15:25:39 HOT-1	gear up.		
		15:25:45 TWR	Cactus fifteen forty nine contact New York departure, good day.

Intra-Cockpit Communication		Air-Ground Communication	
		15:25:48 RDO-1	good day.
15:25:49 HOT-2	heading select please.		
		15:25:51.2 RDO-1	Cactus fifteen forty nine, seven hundred, climbing five thousand.
		15:26:00 DEP	Cactus fifteen forty nine New York departure radar contact, climb and maintain one five thousand.
15:26:02 CAM	[sound similar to decrease in engine noise/speed]		
		15:26:03.9 RDO-1	maintain one five thousand Cactus fifteen forty nine.
15:26:07 HOT-1	fifteen.		
15:26:08 HOT-2	fifteen. climb.		
15:26:10 HOT-1	climb set.		
15:26:16 HOT-2	and flaps one please.		
15:26:17 HOT-1	flaps one.		

Intra-Cockpit Communication		Air-Ground Communication
15:26:37 HOT-1	uh what a view of the Hudson today.	
15:26:42 HOT-2	yeah.	
15:26:52 HOT-2	flaps up please, after takeoff checklist.	
15:26:54 HOT-1	flaps up.	
15:27:07 HOT-1	after takeoff checklist complete.	
15:27:10.4 HOT-1	birds.	
15:27:11 HOT-2	whoa.	
15:27:11.4 CAM	[sound of thump/ thud(s) followed by shuddering sound]	
15:27:12 HOT-2	oh #.	
15:27:13 HOT-1	oh yeah.	
15:27:13 CAM	[sound similar to decrease in engine noise/frequency begins]	
15:27:14 HOT-2	uh oh.	
15:27:15 HOT-1	we got one rol-both of 'em rolling back.	

Intra-Cockpit Communication		Air-Ground Communication
15:27:18 CAM	[rumbling sound begins and continues until approximately 15:28:08]	
15:27:18.5 HOT-1	ignition, start.	
15:27:21.3 HOT-1	I'm starting the APU.	
15:27:22.4 FWC	[sound of single chime]	
15:27:23.2 HOT-1	my aircraft.	
15:27:24 HOT-2	your aircraft.	
15:27:24.4 FWC	[sound of single chime]	
15:27:25 CAM	[sound similar to electrical noise from engine igniters begins]	
15:27:26.5 FWC	priority left. [auto callout from the FWC. this occurs when the sidestick priority button is activated on the Captain's sidestick]	
15:27:26.5 FWC	[sound of single chime]	

Intra-Cockpit Communication		**Air-Ground Communication**	
15:27:28 CAM	[sound similar to electrical noise from engine igniters ends]		
15:27:28 HOT-1	get the QRH . . . [Quick Reference Handbook] loss of thrust on both engines.		
15:27:30 FWC	[sound of single chime begins and repeats at approximately 5.7 second intervals until 15:27:59]		
		15:27:32:9 RDO-1	mayday mayday mayday. uh this is uh Cactus fifteen thirty nine hit birds, we've lost thrust (in/on) both engines we're turning back towards LaGuardia.
		15:27:42 DEP	ok uh, you need to return to LaGuardia? turn left heading of uh two two zero.
15:27:43 CAM	[sound similar to electrical noise from engine igniters begins]		

Intra-Cockpit Communication		Air-Ground Communication	
15:27:44 FWC	[sound of single chime, between the single chimes at 5.7 second intervals]		
		15:27:46 RDO-1	two two zero.
15:27:50 HOT-2	if fuel remaining, engine mode selector, ignition.★ ignition.		
15:27:54 HOT-1	ignition.		
15:27:55 HOT-2	thrust levers confirm idle.		
15:27:58 HOT-1	idle.		
15:28:02 HOT-2	airspeed optimum relight. three hundred knots. we don't have that.		
15:28:03 FWC	[sound of single chime]		
15:28:05 HOT-1	we don't.		
		15:28:05 DEP	Cactus fifteen twenty nine, if we can get it for you do you want to try to land runway one three?
15:28:05 CAM-2	if three nineteen-		

Intra-Cockpit Communication		**Air-Ground Communication**	
		15:28:10.6 RDO-1	we're unable. we may end up in the Hudson.
15:28:14 HOT-2	emergency electrical power . . . emergency generator not online.		
15:28:18 CAM	[sound similar to electrical noise from engine igniters ends]		
15:28:19 HOT-1	(it's/is) online.		
15:28:21 HOT-2	ATC notify. squawk seventy seven hundred.		
15:28:25 HOT-1	yeah. the left one's coming back up a little bit.		
15:28:30 HOT-2	distress message, transmit. we did.		
		15:28:31 DEP	arright Cactus fifteen forty nine its gonna be left traffic for runway three one.
		15:28:35 RDO-1	unable.
15:28:36 TCAS	traffic traffic.		

Intra-Cockpit Communication		Air-Ground Communication	
		15:28:36 DEP	okay, what do you need to land?
15:28:37 HOT-2	(he wants us) to come in and land on one three . . . for whatever.		
15:28:45 PWS	go around. windshear ahead.		
15:28:45 HOT-2	FAC [Flight Augmentation Computer] one off, then on.		
		15:28:46 DEP	Cactus fifteen (twenty) nine runway four's available if you wanna make left traffic to runway four.
		15:28:49.9 RDO-1	I'm not sure we can make any runway. uh what's over to our right anything in New Jersey maybe Teterboro?
		15:28:55 DEP	ok yeah, off your right side is Teterboro airport.
15:28:59 TCAS	monitor vertical speed.		

Intra-Cockpit Communication		**Air-Ground Communication**	
15:29:00 HOT-2	no relight after thirty seconds, engine master one and two confirm-		
		15:29:02 DEP	you wanna try and go to Teterboro?
		15:29:03 RDO-1	yes.
15:29:05 TCAS	clear of conflict.		
15:29:07 HOT-2	-off.		
15:29:07 HOT-1	off.		
15:29:10 HOT-2	wait thirty seconds.		
15:29:11 PA-1	this is the Captain brace for impact.		
15:29:14.9 GPWS	one thousand.		
15:29:16 HOT-2	engine master two, back on.		
15:29:18 HOT-1	back on.		
15:29:19 HOT-2	on.		
		15:29:21 DEP	Cactus fifteen twenty nine turn right two eight zero, you can land runway one at Teterboro.

Intra-Cockpit Communication		Air-Ground Communication	
15:29:21 CAM-2	is that all the power you got? * (wanna) number one? or we got power on number one.		
		15:29:25 RDO-1	we can't do it.
15:29:26 HOT-1	go ahead, try number one.		
		15:29:27 DEP	kay which runway would you like at Teterboro?
15:29:27 FWC	[sound of continuous repetitive chime for 9.6 seconds]		
		15:29:28 RDO-1	we're gonna be in the Hudson.
		15:29:33 DEP	I'm sorry say again Cactus?
15:29:36 HOT-2	I put it back on.		
15:29:37 FWC	[sound of continuous repetitive chime for 37.4 seconds]		
15:29:37 HOT-1	ok put it back on . . . put it back on.		
15:29:37 GPWS	too low. terrain.		

Intra–Cockpit Communication	Air–Ground Communication
15:29:41 **GPWS** too low. terrain.	
15:29:43 **GPWS** too low. terrain.	
15:29:44 **HOT-2** no relight.	
15:29:45.4 **HOT-1** ok lets go put the flaps out, put the flaps out.	
15:29:45 **EGPWS** caution terrain.	
15:29:48 **EGPWS** caution terrain.	
15:29:48 **HOT-2** flaps out?	
15:29:49 **EGPWS** terrain terrain. pull up. pull up.	
	15:29:51 **DEP** Cactus uh. . . .
	15:29:53 **DEP** Cactus fifteen forty nine radar contact is lost you also got Newark airport off your two o'clock in about seven miles.
15:29:55 **EGPWS** pull up. pull up. pull up. pull up. pull up. pull up.	
15:30:01 **HOT-2** got flaps out.	

Intra-Cockpit Communication		Air-Ground Communication	
15:30:03 HOT-2	two hundred fifty feet in the air.		
15:30:04 GPWS	too low. terrain.		
15:30:06 GPWS	too low. gear.		
15:30:06 CAM-2	hundred and seventy knots.		
15:30:09 CAM-2	got no power on either one? try the other one.		
		15:30:09 4718	two one zero uh forty seven eighteen. I think he said he's goin in the Hudson.
15:30:11 HOT-1	try the other one.		
15:30:13 EGPWS	caution terrain.		
		15:30:14 DEP	Cactus fifteen twenty nine uh, you still on?
15:30:15 FWC	[sound of continuous repetitive chime begins and continues to end of recording]		
15:30:15 EGPWS	caution terrain.		
15:30:16 HOT-2	hundred and fifty knots.		

Intra-Cockpit Communication		**Air-Ground Communication**	
15:30:17 HOT-2	got flaps two, you want more?		
15:30:19 HOT-1	no lets stay at two.		
15:30:21 HOT-1	got any ideas?		
		15:30:22 DEP	Cactus fifteen twenty nine if you can uh. . . . you got uh runway uh two nine available at Newark it'll be two o'clock and seven miles.
15:30:23 EGPWS	caution terrain.		
15:30:23 CAM-2	actually not.		
15:30:24 EGPWS	terrain terrain. pull up. pull up. ["pull up" repeats until the end of the recording]		
15:30:38 HOT-1	we're gonna brace.		
15:30:38 HOT-2	★ ★ switch?		
15:30:40 HOT-1	yes.		
15:30:41.1 GPWS	(fifty or thirty)		
15:30:42 FWC	retard.		
15:30:43.7	[End of Recording]		
15:30:43.7	[End of Transcript]		

HARPER LUXE

THE NEW LUXURY IN READING

We hope you enjoyed reading
our new, comfortable print size and found it
an experience you would like to repeat.

Well – you're in luck!

HarperLuxe offers the finest in fiction and
nonfiction books in this same larger print size and
paperback format. Light and easy to read, HarperLuxe
paperbacks are for book lovers who want to see
what they are reading without the strain.

For a full listing of titles and
new releases to come, please visit our website:

www.HarperLuxe.com